CITY OF CHANGE AND CHALLENGE

City of Change and Challenge

Urban Planning and Regeneration in Liverpool

CHRIS COUCH
Liverpool John Moores University

Routledge
Taylor & Francis Group

LONDON AND NEW YORK

First published 2003 by Ashgate Publishing

Reissued 2018 by Routledge
2 Park Square, Milton Park, Abingdon, Oxon OX14 4RN
711 Third Avenue, New York, NY 10017, USA

Routledge is an imprint of the Taylor & Francis Group, an informa business

Publisher's Note
The publisher has gone to great lengths to ensure the quality of this reprint but points out that some imperfections in the original copies may be apparent.

Disclaimer
The publisher has made every effort to trace copyright holders and welcomes correspondence from those they have been unable to contact.

A Library of Congress record exists under LC control number: 2003048921

ISBN 13: 978-1-138-71595-0 (hbk)
ISBN 13: 978-1-138-71594-3 (pbk)
ISBN 13: 978-1-315-19726-5 (ebk)

Contents

List of Figures and Tables

Figures

Tables

Preface

I first visited Liverpool when I was eight years old. We took a trip on the 'overhead railway' that, in those days, ran alongside the docks from the Dingle to Seaforth. All I can remember is an impression of miles and miles of docks containing lots and lots of ships. I came from Bristol, itself a significant port, but I had never before seen so many ships and so much activity. It was very impressive.

My second visit to Liverpool was some sixteen years later in 1971. We arrived at night, driving in along the A57 from Warrington. The built up area started near Prescot, more than nine miles from the Pier Head. We followed the brightly lit, broad dual-carriageway of Liverpool Road, then East Prescot Road past endless inter-war housing estates and shopping parades, across Queens Drive, past Old Swan and into the 19th century city through Fairfield and Kensington and finally over the ridge and down London Road and into the centre. This was a big city. The area around Liverpool's magnificent 18th century town hall contained commercial and financial offices of truly metropolitan scale. This was a serious business district - grander than anything I had seen in any English city outside London. The city centre had seven department stores (yes seven - Hendersons, G H Lee's, Lewis's, Owen Owen, Blacklers, T J Hughes and the Co-op); three mainline railway termini - Lime Street Station, Central Station and Exchange Station; and one of the biggest municipal museums and art galleries in the country. It had a cultural scene of national importance and two top football teams - one of which was the best in Europe. This was an important city.

We did not know it at the time but this was the end of an era. Everything was about to change. Competition, recession and political failure did for that city. Today things are very different: the city is still big: but smaller than it was. Suburbanisation and outward migration have robbed the city of a third of its population in less than 40 years. The port employs only a fraction of the huge workforce that used to handle cargoes from every corner of the globe. Economic restructuring and international competition have decimated manufacturing industry. The commercial importance of the city centre has declined sharply, especially in the face of competition from other centres in the region. The overhead railway is long gone; much of the port has closed; important firms have left the area and office employment has declined. Only three department stores remain. On the other hand, for most people, living and working conditions have improved. The main shopping streets are pedestrianised and there is a new fashion for city living and a revived interest in culture. Many of the former docklands have been transformed into residential and commercial areas, hotels, marinas and galleries.

The title of the book is taken from a marketing epithet used by the former Liverpool Corporation in the early 1970s. The challenge to the Council and others has been to respond to these changes and to plan its future development. This planning agenda has evolved over time. In the 1960s the main concern was to modernise the city: to provide decent homes for all, schools, shops and open spaces; and to provide a modern infrastructure that could accommodate the rising demands of road traffic. During the 1970s and 1980s as the city faced recession, regeneration became the urgent challenge. The need now was to find new firms to replace those that had closed, to create new jobs to replace those that had been lost, to reclaim land that had been laid derelict through industrial change and to improve the lot of people suffering social deprivation. By the 1990s new challenges were being added to the agenda: the need to compete with other regional cities in Britain and even across Europe for footloose investment and development; and the need to create a city that was environmentally sustainable; healthy and socially inclusive.

This book is about those changes and challenges. It is about urban change, urban planning and regeneration over the last 40 years. While the focus of the book is the city of Liverpool many of the changes, trends and responses found here can also be found elsewhere in other British cities and beyond. For whilst Liverpool may represent an extreme case in terms of the rate of urban change, the nature of the pressures exerted upon its economy, society and environment are no different from those found in other cities. The lessons to be learned from a study of Liverpool can be widely applied in many other British and European cities.

One of my colleagues, who was kind enough to read an earlier draft of this text, commented critically that in a city supposedly so full of 'characters', there is little reference to how people have helped shape the events or outcomes; everything seems to be process driven without 'human agency'. I did not think this was an entirely fair comment as the book does identify a number of key actors at various stages in the process. Nevertheless, there is something in this point. I have not made great play of the personalities involved and I accept that I may be criticised for this. However, it was (at least in part) my intention not to dwell on the roles of individuals. In my view economic and social forces and processes have been much more important in shaping the fortunes of the city and the nature of policy responses than the charisma, skills or leadership of any individual. Nevertheless, I accept that on occasions the actions of certain people may have modified or shaped the detail of events or policies.

I was a town planner working in the area during the 1970s and have spent the last 25 years teaching and researching urban planning and regeneration at Liverpool John Moores University. I have therefore been a close witness to many of the changes and much of the policy responses in Liverpool. This proximity has given me access to documents and conversations with many people. All this has provided a rich source of background material and I am grateful to the many individuals who have provided the, occasionally unwitting, source of ideas in different parts of the book. Not least among these have been my excellent colleagues in the Planning and Housing Section at JMU. In particular I would like to thank David Alexander and Steven Fowles for their helpful comments on earlier drafts of this text and Paul Hodgkinson for his first class work on the maps and illustrations. However, I must take responsibility for the contents of this final text, errors and all. Like much academic work, this book was largely written in the vacations, evenings and weekends and my final thanks go to my wife, Lynda, for her forbearance in putting up with this tiresome process.

Chris Couch
Liverpool John Moores University

1 Introduction

The Changing Nature of Urban Planning

Urban planning in Britain has changed and developed over the last 40 years from being an adjunct of architecture and municipal engineering to emerge as a powerful professional discipline in its own right, albeit one that seems to suffer all to frequent crises of identity and confidence. As planning decisions directly affect the wishes of land owners, property values and the profitability of urban development it is hardly surprising to find it under near constant attack from a wide variety of vested interests. As the planning system has evolved so the debates surrounding its purpose, costs and efficiency have intensified.

In the early 1960s planning operated at a number of scales from the preparation of county-wide development plans, through the production of town maps, to plans for the comprehensive development or redevelopment of small areas. But whatever the scale, the role of the planner was primarily concerned with land use zoning and resolution of physical problems associated with development. However, through that decade there was something of a revolution in the nature of town planning. First came recognition of the growing problem of traffic in towns and following the publication of the Buchanan Report (Ministry of Housing and Local Government, 1963), the need for an integrated approach to land use and transportation planning. There was also a growing reaction against modernist approaches to architecture and planning. Following the pioneering work of Ian Nairn and others in the 1950s, the early 1960s saw the publication of books such as Gordon Cullen's *Townscape* (Cullen, 1961) and Jane Jacobs' *Death and Life of Great American Cities*, (Jacobs, 1964). Such influences spearheaded the development of more sensitive approaches to urban planning and an emerging acceptance of conservation and heritage protection as matters of legitimate planning concern. Towards the end of the decade, coinciding with an influx of social scientists into the profession, came a growing concern to plan the city in its social and economic dimensions as well as its physical form.

By the early 1970s the 'inner city' and 'urban deprivation' had emerged as planning problems and after the post-1973 recession, local economic development became a major issue. The need to tackle these and other problems such as urban dereliction, outworn infrastructure and under investment in inner city housing led to the notion of urban regeneration as an important aim of urban planning. By the late 1970s radical social commentators were raising questions from feminist, ethnic minority and environmental perspectives that challenged many of the traditional assumptions and values of planning. Margaret Thatcher's ministers questioned the whole validity of the planning process and the need for intervention in market-led development processes. Her government's approach to urban regeneration was characterised by a close association with the private sector and an emphasis on property development. In 1987 the Brundtland Commission Report *Our Common Future* (World Commission on Environment and Development) defined the concept of 'sustainable development', an idea that was given additional support by the 1992 *UN Conference on Environment and Development* (The Rio Summit) as well as various European Commission and UK government policy statements. Thus by the early 1990s environmental protection and resource conservation had moved from being fringe issues of minority interest to matters of mainstream concern, in which town planning, urban regeneration and urban design could make a significant contribution. Thus the profession that was in the early 1960s limited to land use and physical planning, was by the millennium also engaged with transportation policy, urban conservation, urban social policies, regeneration, economic development and environmental sustainability.

Whilst the planning agenda has expanded hugely over this period so the nature of the planning process is also said to have changed. Many writers suggest that planning has moved through a series of phases (Healey, 1997; Rydin, 1998; Taylor, 1998). Until the mid 1960s it is said that planning was characterised by an emphasis on the production of 'blueprint plans' or 'master plans' in a direct reference to the design processes used in architecture and engineering. To put it crudely, town planning was viewed as little more than architectural design on a larger scale. The late 1960s saw the growth of a perspective that envisaged town planning as an ongoing process of 'urban management' rather than one that stopped with the production and implementation of a plan for physical development. Cities became viewed as 'systems' that were in a state of perpetual change (Mcloughlin, 1969; Chadwick, 1971). The role of town planning became more concerned with the management and control of urban change to meet certain goals, be they a balance between housing supply and demand, the conservation of areas of architectural or historic interest, or whatever.

It gradually became clear that the definition and prioritisation of goals for urban development were matters of significant political concern. With this recognition came a new perception of planning as a part of the political process rather than simply a technical activity. As the availability of resources declined during the recession of the 1970s, especially in the public sector, so choices between conflicting goals became harder and harsher. Planning decisions became more controversial as more individuals and more groups found themselves on the losing end of the processes that allocated these increasingly scarce resources. For a time the planning profession was seen to side with the radical ideologies of the New Left against Thatcherism. But others saw planning as little more than a component part of the development process, supporting the mantra of 'property-led development' to create 'much needed jobs' (See Rydin, 1998, for a discussion). In reality planners were probably more technocratic and apolitical than this and in pursuing their traditional concerns for 'amenity' and 'efficiency' in urban form found themselves caught in the middle ground between conflicting interests. Nevertheless, the role of the planner gradually moved away from that of technical expert into that of communicator and negotiator, mediating between the conflicting interests of different groups, usually with developers on one side and conservationists or local communities on the other. The result has been an unpopular profession, for in these disputes everybody loses something and everybody blames the messenger.

Changes in Liverpool

Over the last 40 years Liverpool has undergone more economic restructuring and urban change than virtually any other city in Britain or Europe. It has lost 40 per cent of its population and more than half its manufacturing employment. The experience of this city can teach us a great deal about the nature of urban change and the policy responses of central and local government.

Liverpool has been a laboratory for almost every experiment and innovation in modern urban policy and planning. Between 1967 and 2000 the city played host to a Community Development Project; Education Priority Area; Inner Areas Study; Area Management; Industrial and Commercial Improvement Areas; one of the first General Improvement Areas in the country; more Housing Action Areas than any other city; an Inner Area Partnership; a Metropolitan Structure Plan; the Merseyside Development Corporation; more housing co-operatives than any other English city; an Enterprise Zone; a 'Minister for Merseyside'; a 'militant' Labour Council at war with a Tory Government; Estate Action; City Challenge; a Housing Action Trust; Single Regeneration Budget projects, an urban regeneration company and many, many more initiatives generated by central government, local government and other agencies from the public, private and voluntary sectors.

Urban Planning and Regeneration in Liverpool

The purpose of this book is to describe and examine the changing nature of urban planning in Liverpool over this critical period from the early 1960s to the millennium. The focus of interest is on the content and nature of plans and policies rather than the socio-political process by which policy came about. The scope of the book is also limited to urban planning and particularly the regeneration agenda that emerged from the early seventies. Thus little is said here about planning problems at the periphery of the conurbation, the green belt or rural issues. On the other hand, the city centre and the inner city are of particular interest.

What is offered here is a description of urban planning and regeneration in Liverpool over the last 40 years. This experience is used as a vehicle for a discussion of a number of issues including the extent to which:

- urban regeneration has moved from being a novel concern of urban planning to gradually dominate and ultimately to usurp the planning process;
- urban planning and regeneration policies have failed to improve the relative position of deprived parts of the city *vis-à-vis* the city average, or the relative position of the city within the national context;
- the economic development aims of policy makers have consistently taken precedence over environmental and social concerns;
- urban planning and regeneration have been dominated by top-down approaches, with increasingly centralised decision making, limited local participation, democracy and accountability;
- the regeneration and development process has been ad hoc and unplanned, whilst much urban planning and regeneration policy has been reactive and lacking in strategy or long term vision.

These issues are explored within each chapter and debated within the concluding chapters. The structure of the book attempts to be both chronological and thematic, discussing issues at points in time when they perhaps first became important or were the subject of significant discussion.

Chapter two gives an impression of the extent of change in the city, especially over the last 40 years. From the early days of the port, the economic fortunes of the city are outlined. Whilst the process of economic restructuring can be traced back through decades, the early 1970s saw a major downturn in the prosperity of the city. Unemployment rose sharply and continued at a high level throughout much of the next two decades, having a dramatic impact on the local community and playing a major part in shaping economic and social policy. Partly as a consequence of economic decline and partly as a result of local housing policy and demographic trends, population decline has been a major feature of the city throughout the period. Not only has the population of the city fallen sharply since the sixties but in comparison with other areas, the residual population is now relatively older, poorer and less skilled than it was four decades ago. This has led to increased stress for individuals and increased pressure on local health, education and social services. In contrast the housing situation has eased considerably, especially in quantitative terms. However, the present crude housing surplus is at the root of many of today's problems of hard-to-let and abandoned dwellings in unpopular areas. The period since 1971 has seen major shifts in tenure and the commodification' of housing. Despite large sums invested in housing renewal, much of the city's stock is of poor quality and in need of substantial repairs. But then housing is no longer the important political issue in Britain that it was in the post-war years.

The environment of the city has also changed. Much of the physical infrastructure of the city has been renewed or refurbished. The industrial restructuring of the 1970s and 1980s left a legacy of vacancy and dereliction, although by the millennium much of this land had been redeveloped and reused in one way or another, albeit often at a much lower density. Great progress has been made in the preservation and conservation of buildings and areas of architectural or historic interest, although much of the physical environment elsewhere is poorly managed and maintained. Pollution from industrial sources has declined sharply whereas pollution from road traffic has increased. Traffic has become one of the key sources of serious environmental degradation affecting in many parts of the city. In comparison with many other European cities the environment of Liverpool remains relatively poor.

It was in the mid 1960s that the first modern plan for Liverpool was produced. This period, discussed in chapter three, was characterised by a number of substantial developments in town planning. Nearly two decades of experience with preparing and implementing development plans under the Town and Country Planning Act 1947 had led many to call for changes in the planning system. In response to such calls, the government established a Planning Advisory Group on *The Future of Development Plans*. The group reported in 1965 and advocated a broadening of the planning agenda to include traffic management and protection of the environment (Ministry of Housing and Local Government, 1965). They also called for the separation of strategic planning (general policies and major proposals) from tactical planning (detailed land use and development decisions). These proposals passed into legislation with the Town and Country Planning Act 1968. The Interim Planning Policy Statement (IPPS) produced by Liverpool City Council in 1965 was in many ways a prototype for these new strategic plans, or 'structure plans' and it is no coincidence that Walter Bor (the City Planning Officer) was an influential member of the Planning Advisory Group. The predominant aim of the IPPS was to remodel the physical structure of Liverpool into a modern city that would develop around efficient transport systems closely integrated with planned changes in land use. But changes in the city's fortunes and changing fashions in planning made implementation difficult and much of that ambition remains unfulfilled even today.

Gradually planning was beginning to embrace the broader social and economic problems that were becoming apparent in many urban areas. Liverpool was at the forefront of this movement. During the late 1960s the City Council undertook a series of social surveys which, together with analysis of the new 1966 Census and other data, provided one of the first comprehensive studies of 'social malaise' to be undertaken in a British city. The distribution and intensity of social and economic problems were carefully mapped as the basis for a new kind of more comprehensive urban policy intervention aimed not only at improving physical conditions but also social and economic aspects of residents lives. By 1970 the Council had divided the city up into three zones for planning purposes and produced separate plans for the city centre, the inner areas and the outer areas. Thus, for the first time, the inner areas became recognised as a part of the city that had particular problems and required their own specific type of policy interventions.

Chapter four is concerned with the early studies, experiments and policy responses. It was in 1968 that the Home Office launched what many would regard as the first truly modern intervention in urban policy: the Urban Programme. Within a couple of years the Home Office had decided that an action-research programme was needed to further explore the nature of urban deprivation and the role of community development. One of these so-called Community Development Projects (CDP) was established in the Vauxhall area of Liverpool. Shelter, the national charity for the homeless, was also keen to investigate solutions to housing deprivation and set up the Shelter Neighbourhood Action Project in the Granby district in 1969. The Home Office was not the only government department concerned about the emerging problems in inner urban areas: the Department of the Environment was developing its own strategies. In 1972 it appointed consultants to research in three case study authorities: Birmingham, Lambeth and Liverpool. Later, the findings and lessons from this work would inform the content of the Labour government's Inner Urban Areas Act 1978. Thus the early 1970s were an important period in the development of new forms of urban policy with Liverpool, through experiment and experience, very much to the fore.

Chapter five considers the contribution of development plans during this period. By 1970 the government had established its recently modernised and comprehensive planning system. The government itself was preparing plans for each of the English regions. Local planning authorities were beginning to prepare county and urban structure plans in order to provide a strategic framework for more detailed local planning and development. But almost as soon as this system was in place it was being undermined by changing economic circumstances and by the government's own actions. There had already been a balance of payments crisis in 1968 that had led to cutbacks in public spending, including urban road building and housing schemes. The deep recession of the mid 1970s led to further swingeing cuts in the budgets of public authorities. Capital expenditure on infrastructure, public transport and housing were particularly affected as well as local authority revenue expenditure on building maintenance and repair and the delivery of mainstream services. At the same time the reorganisation of local government under the Local Government Act 1972 split the planning system, with county councils now responsible for structure plans and some local plans and the new district councils responsible for most local plans and development control. Further, many of the emerging urban regeneration programmes, such as housing improvements and local economic development initiatives were being implemented in an *ad hoc* manner outside any formal planning context. Nevertheless, despite all of these problems the decade saw the production of two key strategic planning documents that affected the conurbation: the Strategic Plan for the North West and the Merseyside Structure Plan. Both had their origins in the planning system established by the Labour government in the late 1960s and provided important examples of inter-corporate working towards shared aims. Both made significant contributions towards a broadening of the planning agenda, especially in the fields of urban regeneration, environmental protection and natural resource management. But both were completed at a time when planning and urban policy had become more piecemeal and tactical and where there was now little scope for intervention between the level of national policy guidelines and the implementation of local programmes and projects.

Chapter six explores urban regeneration during the Thatcher era. This was a time of little strategic planning but many uncoordinated programmes for re-investment in urban areas. During the mid-seventies the Labour government worked closely with local government on the issue of urban deprivation and by 1978 had formulated its policy and passed the Inner Urban Areas Act. Partnership between the different levels of government was a key feature of their strategy, together with a redirection of funding towards hard-pressed inner city authorities. However, with the election of the new Conservative government in 1979 the agenda swiftly changed to emphasise a much greater concern to promote 'property-led' regeneration strategies for run-down urban areas. Furthermore the Conservatives mistrusted the ability of local authorities to deliver on this agenda and moved away from the central-local partnership model to one that relied more directly on central government working with the private sector to deliver urban regeneration. This was the era that saw the designation of the Merseyside Development Corporation and the Speke Enterprise Zone. Riots in Toxteth and elsewhere led to policy modifications and saw Michael Heseltine appointed, for a short time at least, as 'Minister for Merseyside'. Out of this initiative came the influential Merseyside Task Force. Through the nineteen eighties central government constantly seemed to be seeking ways of streamlining (some would say sidelining) the planning system and centralising more and more regeneration decisions to themselves.

Chapter seven considers the evolution of planning policies that culminated in the Liverpool Unitary Development Plan 1996. In 1983 the City Council produced a statement of 'Current Planning Policies and Development Programmes: Liverpool'. This was not a formal development plan, indeed the Council denied that it was a plan at all. However, in elaborating the policies of the Structure Plan and providing a framework for local planning activity it usefully filled the gap between strategic policy and the tactics of day to day development control. Nevertheless, in general the 1980s were characterised by central government scepticism over the value of local development plans, a centralisation of planning powers and a generally more relaxed and permissive approach to the control of property development.

By 1986 the Merseyside County Council had been abolished and in place of the Structure Plan Liverpool City Council was required to produce a 'Unitary Development Plan' (UDP), incorporating statements of general development strategy as well as detailed proposals and policies for the control of development. The Liverpool UDP was published in draft form in 1996 but delays meant that a final document had still not been formally approved even at the end of the millennium.

Chapter eight discusses regeneration in the 1990s. By the end of the 1980s the government were being roundly criticised for the way local authorities had become marginalised in the regeneration process, for the complexity of their regeneration policies and the proliferation of agencies and sources of funding (Audit Commission, 1989). With the return of Michael Heseltine to his old post in 1991, local authorities were brought back into the process as the local leaders of bids for 'City Challenge' funding. In 1993 John Major's new Conservative government also set about simplifying the regeneration process by merging more than twenty different funding streams into one 'Single Regeneration Budget' (SRB) whilst at the same time bringing additional co-ordination through the establishment of Regional Offices of Government and a new regeneration agency for England (English Partnerships). Liverpool benefited from a successful City Challenge bid, a number of SRB schemes and various investments by English Partnerships. However, despite this fairly ambitious attempt to simplify and bring more co-ordination to urban planning and regeneration it was soon clear that the system was not working as well as it might. With a growing number of, often competing, agencies at work and the slow delivery of development plans, in many areas the regeneration process was disintegrating into a series of localised projects that were lacking in strategic context and focused only on short-term goals. This process continued under the new Labour government after 1997 and despite rhetoric to the contrary, regeneration in the nineteen nineties became increasingly characterised by fragmentation and a lack of co-ordinated planning.

Chapter nine considers what has been achieved in by urban planning and regeneration policies in Liverpool over the past four decades. It is concluded that there has been a general failure to improve the relative position of deprived parts of the city *vis-à-vis* the city average, or the relative position of the city within the national context. A reflective look back over the changes in the policies and delivery of urban planning and urban regeneration in the city suggests that there is considerable evidence that economic development has consistently taken precedence over environmental and social concerns. With notable exceptions, policy-making has generally been dominated by top-down approaches, with increasingly centralised decision making, limited local participation, democracy and accountability, although there have been significant variations over time.

Finally, it is suggested that that the achievement of urban regeneration, however ill defined as a concept, has become more important to politicians than the achievement of a well-planned environment. Over recent decades, urban planning has become increasingly reactive and lacking in long term visions for the future development of the city. At the same time much of the regeneration activity has been ad hoc and unrelated to any strategic plan. Nevertheless, it seems probable that the fortunes of the city will stabilise as it finds its place in the new urban hierarchy.

2   City of Change

The Port

The history of the port of Liverpool is well known and will not be repeated here. Suffice to say that the origins of the modern port can be found in the growth of colonial trade and the symbiotic relationship between Liverpool and the growing industrial towns of Lancashire and Cheshire in the mid 18th century.   This burgeoning trade stimulated investment in the infrastructure of transport and there was a need for more quaysides, warehousing and other port-related development. This demand, together with the problems of tidal range on the Mersey, led to the building of the first dock in Liverpool, at the 'pool' in 1715, a second dock (Salthouse Dock) followed in 1759. Subsequently there was a rapid investment with St Georges Dock opening in 1771, the Kings Dock in 1788 and the Queens Dock in 1799. The Liverpool to Manchester railway, the first railway built to carry passengers as well as freight, opened in 1830. The Bridgewater Canal, linking Manchester with Runcorn on the Mersey, opened in 1767 and the Leeds and Liverpool Canal in 1774. A century later virtually the whole of the south dock system was complete and most of the northern docks as far as Huskisson Dock were in place. Complementing the docks were shipping offices, brokers and all manner of related industrial and commercial activities (Marriner S, 1982, Ch 2 and 3).

Such massive infrastructure investment stimulated yet further growth in trade. The total net registered tonnage of ships using the docks rose to 4.4 million tons by 1858, 9.6 million tons by 1890 and 19.0 million tons by 1914 (Hyde F, 1971, p96). By the latter half of the 19th century the good fortunes of the port had turned Liverpool into a prosperous and thriving metropolis. The trappings of this wealth could be seen in the scale of the docks themselves, the opulence of the commercial buildings that housed the shipping companies, commodity exchanges, banking houses and insurance companies that grew up alongside the shipping trade. Large sums were also being spent on civic buildings, public works, parks and gardens. The middle classes built themselves grand terraces and mansions, firstly in the Georgian and early Victorian terraces and squares around Rodney Street, Canning Street and Abercrombie Square and later in more distant suburbs such as Toxteth, Princes Park and Sefton Park. Only the working classes remained relatively unrewarded in an inequitable distribution of trading profits. For them Liverpool was frequently a city of casual and poorly paid work, slum housing and bad health.

The first signs of decline in the fortunes of the city were evident from the beginning of the 20th century, although somewhat masked by the key role played by the port in both world wars. Despite the period before the World War I being the heyday of passenger traffic, Liverpool's share of UK emigrant traffic was already dropping. By 1907 the White Star Line had transferred its express passenger services to Southampton and Cunard, complaining about passenger facilities, similarly transferred its large liners after the end of World War 1 (Marriner S, 1982, p96). By the end of the 1960s deep-sea passenger liners had deserted the port completely, leaving only cross-river traffic and ferry services to Ireland and the Isle of Man. The port of Liverpool faces the Irish Sea and the Atlantic Ocean. In recent times, with the relative decline in colonial and deep sea traffic and the growing importance of European trade, Liverpool found itself to be on the wrong side of the country and increasingly uncompetitive against the ports of south-east England: from Southampton to Felixstowe. Between 1966 and 1979 Liverpool's share of UK short-sea trade with Europe fell from 6.1 per cent to 2.4 per cent and its share of deep-sea trade fell from 24.5 per cent to 13.8 percent (Gilman S and Burn S, 1982, table 3.1).

The methods of dock working changed little during the 19th and early 20th centuries so that the only way to accommodate increases in trade had been through the expansion of the dock system and dock estate. When technological change did occur it arrived with great speed and dramatic impact. During the 1960s and early 1970s new methods of cargo handling were introduced. Combined with the trend towards larger vessels, the effect was to increase speed of cargo handling and reduce the demand for wharfs, especially in the older, small docks to be found south of the Pier Head. Throughput per berth rose from 50,000 tonnes per annum, to between 400,000 and 1,500,000 tonnes per annum for a single link span. Even if growth had continued, the growing use of container ships and roll-on roll-off ferries would have led inevitably to a reduction in the size of the dock estate (Gilman S & Burn S, 1982, p27). A further effect of the new transport technologies was a reduction in labour demand. A typical conventional berth operating with three gangs could handle about 250 tonnes a shift; the same number employed in two container gangs could move more than 2500 tonnes.

The Port of Liverpool responded to changes in transport technology and trading patterns with a combination of policies of investment and rationalization. The major investment was in the Royal Seaforth complex...and the main rationalization was the closure of the whole of the

South Docks in 1972. In the rest of the North Docks...there has been a mixture of some smaller scale and planned investments with some areas of low intensity use and pending closure (Gilman and Burns, 1982, p30).

The City Council seems to have been slower than the Mersey Docks and Harbour Company in anticipating change. As late as 1965 the former was still anticipating that there would be pressure for expansion of the northern dock estate and even the southern docks were expected to have to deal with increased traffic (Liverpool City Council, 1965, p91). However, by 1979 the Merseyside County Council was taking a more pessimistic view of the situation.

Although the port will continue to play a major part in the local economy, it is, in the County Council's view, unlikely to return to its former dominance and volume of trade. The future Port of Liverpool will probably be smaller, better equipped, more capital intensive and highly specialised. It is likely to require only limited additional areas of land for port-related uses. Meanwhile, it will be vital for land surplus to port requirements to be made available and attractive for redevelopment (Merseyside County Council, 1979, p23).

If expansion had been a catalyst for the growth of port-related industry and commerce, the decline of the trade through the port coincided with and was partly responsible for their departure in more recent times. Liverpool had been the headquarters location of several major companies but by the nineteen sixties many had moved elsewhere, notably to London (Daniels P W, 1983, p247). Although the port had been the basic industry of Liverpool, the historic foundation of its economic growth and the key to understanding the structure of local employment and commerce, its place in the modern economy was becoming more modest. By 1980 the number of dockworkers had fallen from its 1920 high of around 20,000 to little more than 4,000 (out of a total of over 600,000 jobs in Merseyside as a whole). Docks were becoming abandoned and vast areas laid to waste. In the South Docks alone there were some 90 hectares of derelict land (Stoney, 1983, p121). Although there has been a significant recovery in trade through the port in more recent years, in terms of employment the docks now remain marginal to the economy of the city.

Despite this decline, it should not be forgotten that the docks have central to determining the physical image of the city. The Albert Dock, designed by Jesse Hartley is the largest complex of grade I listed buildings outside London. The three huge office buildings at the Pier Head: Thornley's 1907 renaissance style Port of Liverpool Building, Aubrey

Thomas's 1907 Royal Liver Building and the italianate Cunard Building (1916) are amongst the most important architectural treasures of the city. Together the Pier Head and the Albert Dock are symbolic of the city in its economic heyday and provide the shorthand images by which the city is identified worldwide.

Trends in the Rest of the Local Economy

In 1965 the City Council could still boast that:

> Liverpool is Britain's second port handling 25% by value of the nation's trade. It is a major marketing, industrial, transport and cultural centre, and is of key importance to the social and economic well being of a widespread hinterland, extending from North Wales to the Lake District and from the Irish Sea to the Pennine Uplands (Liverpool City Council, 1965, p23).

Since that time Liverpool, whilst still at the heart of one of the largest conurbations in the country, has suffered a remarkable economic and demographic decline. Although it remains a major port, competition and changes in transport and industrial technology have reduced its importance in the national economy. Today the conurbation, once central to the trading routes between the industrial north of England and the rest of the world, now finds itself on the periphery of an increasingly integrated European economy and reliant of external subsidies to survive.

De-industrialisation brought about by reductions in demand for traditional products and intensifying competition from elsewhere has eliminated much of the industrial base, employment and social stability that existed in the sixties. Competition from other regional centres such as Chester and Warrington and the development of 'out-of-town' shopping and leisure facilities, has reduced the relative importance of Liverpool City Centre in the retail and cultural life of the region. The image of the city is no longer that of a thriving cosmopolitan port but of a place struggling to come to terms with its reduced importance and its poverty.

In the competition for new investment the city has been less successful than some of its competitors. Manchester has become the dominant regional office centre in the North West, aided by its more central location, its international airport and other agglomeration economies. Smaller towns and suburban centres in the region have also benefited more than Liverpool from the growth of small and medium sized firms and the new high technology industries. Nevertheless, Liverpool has not been without its successes in economic development. Culture and tourism, higher education, health and public services have all become important features of the local economy. Some sectors of manufacturing such as motor vehicles, chemicals, pharmaceuticals, food and drink production have also remained important in the wider Merseyside economy.

Even within the relative economic prosperity of the mid-1960s there was a recognised need to attract more growth industry to the area. Male unemployment in Liverpool at that time was only 4.6 per cent and there were shortages of skilled labour in certain fields such as engineering. Figure 2.1 shows subsequent trends in the unemployment rate.

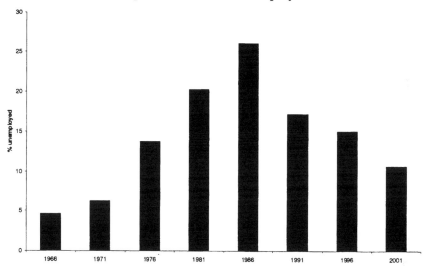

Figure 2.1   **Unemployment rate in Liverpool**

Source: 1966 - 1986   Liverpool City Planning Department
                   1991 - 1996   Calculated from Regional Trends
                   2001   Department for Education and Skills:
                                    Labour Market Statistics, October 2001.

It was also being suggested that the provision of employment was failing to match population growth. Inadequacies in the local transport system, particularly the lack of investment in the road network, were seen as hindrances to economic growth. Thus in the sixties, the main economic problems were seen as a lack of growth industry, skill shortages, employment failing to keep pace with population growth and inefficiencies in the transport system. Nationally there was already in place a powerful set of regional economic development policies. For much of the post-war period the city and much of the conurbation had benefited from

Development Area status and firms had received various forms of subsidy to encourage them to invest in the area. Writing in 1970, Lloyd concluded that during what he termed the 'motor industry phase 1960-7', the impressive growth of vehicle manufacturing, engineering and electrical goods contrasted sharply with the relative decline evident elsewhere (e.g. in shipbuilding, chemicals, textiles, food, drink and tobacco sectors) (Lloyd P E, 1970, p402). Despite some 26,000 new jobs having been created in three new vehicle manufacturing plants (at Halewood and Ellesmere Port) he concluded that:

> the catalytic effects which caused the transfer of labour and closure of many plants in traditional industries were re-working its industrial economy. In the long run the structural changes which took place during these years are likely to prove far more significant than the actual expansion in employment opportunities (Lloyd P E, 1970, p398).

Indeed, as shown in Figure 2.2, over the whole period from 1961 to 1985 only in the vehicle industry did the local economy perform better than the national average. In all other sectors local employment was either declining faster or growing more slowly.

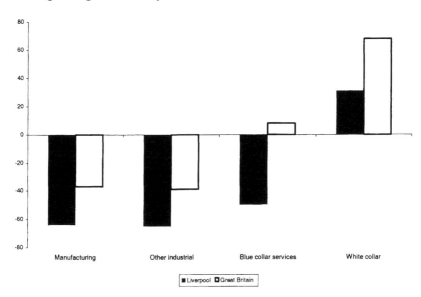

Figure 2.2   Changes in employment structure 1961-85: Liverpool and Great Britain

Source: Table 16, Past Trends and Future Prospects, Liverpool City Council, 1987

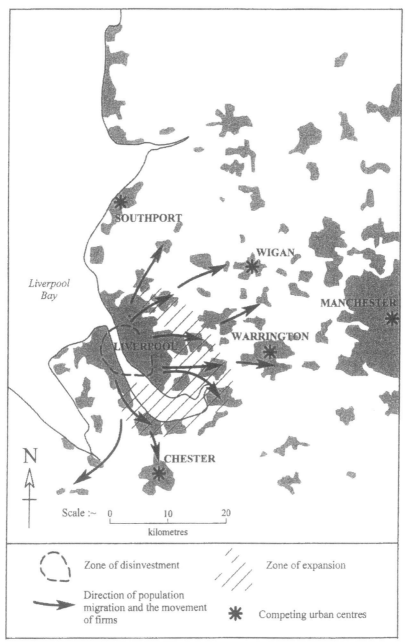

Figure 2.3   The location of investment in and around Liverpool

Source: author

Unfortunately the location of the new job opportunities was not particularly accessible to those being made redundant in the port and inner industrial zones. Most of this new industrial investment was to be found in a peripheral arc running from Kirkby in the north, through Halewood and across the Mersey to Runcorn and Ellesmere Port, whilst decline was concentrated in the inner urban areas. Speculative private housing development was fuelling suburbanisation northwest along the Mersey coast towards Formby and Ainsdale, northwards towards Maghull and Ormskirk, eastwards towards St Helens and Warrington and southwards across the Mersey into the Wirral and even as far as North Wales. Social housing was being provided on overspill estates from Kirkby to Halewood. There were 'expanded town' agreements with Widnes, Winsford and Ellesmere Port and 'new towns' being developed at Skelmersdale, Runcorn and later, at Warrington. All of which was leading to urban sprawl on a massive scale. Figure 2.3 gives a generalised picture of the location of expansion, investment and disinvestment over the period.

In the midst of these changes the City Council concluded that in order to meet the growing employment needs of the population, substantial additional areas of land should be identified for new industrial and service development. Further, they recognised that due to the shortage of land in the city, most of the growth of employment, especially in manufacturing industry, would occur beyond the city boundary. It was also argued that the urban renewal programme would further increase demand for industrial land by displacing many businesses in the older areas and necessitating their relocation elsewhere (Liverpool City Council, 1965, p42).

This approach appeared to start from a projection of population growth that had been made independently from any sound analysis of economic trends. There was an assumption that by providing industrial land the employment needs of the population could be met. As subsequent events were to demonstrate, this approach overestimated the extent to which economic growth could be influenced by public policy, especially by the local authority's development plans, and underestimated the extent to which population would migrate elsewhere in search of work. Secondly, in common with national policy at the time, there was a strong emphasis on meeting the needs of manufacturing industry and relatively little anticipation of the importance of service sector employment in future economic growth. Thirdly, the appearance of great swathes of vacant and derelict land in the inner city had yet to happen. As late as the mid-1960s, most of the docklands were still operational, railway goods yards were still in use, energy suppliers and large manufacturing forms were still major land users in the city, so the periphery was the only place where it was thought possible for expansion to occur. Even urban renewal would force industry outwards as 'non-conforming users' were to be cleared from the inner areas and replaced by new residential developments, schools and shopping parades. Public policy and economic trends were both working in the direction of decentralisation and suburbanisation. The problems of inner city decline had yet to be perceived.

Later, consultants Wilson and Womersley were to comment that the changing structure of local employment had been matched by locational change with substantial losses in the inner areas of the city and gains mainly to be found on the periphery: to the east in Knowsley, St Helens and Halton and to the south in Wirral and Ellesmere Port. In their view firms had left Liverpool because of a lack of room for expansion on existing sites and the economic uncertainties of operating in the inner areas. Peripheral locations offered better availability of premises and land and good accessibility to the motorway network. Even the Government's own development agency, the English Industrial Estates Corporation (EIEC) (subsequently part of English Partnerships) had concentrated its investment on green field sites. Of the 320,000 square metres of industrial space that EIEC developed within Merseyside up to 1976, none lay within the inner areas of Liverpool or Birkenhead (Department of the Environment, 1977, pp 97-102).

By the seventies there was a substantial amount of vacant and derelict land within the inner areas. Port facilities were being rationalised with the abandonment of many older, smaller docks. The emergence of natural gas led to the closure of many 'town' gas works. Older, urban, coal-fired electricity power stations were being replaced by modern, out of town, oil, coal or nuclear power stations. Falling rail traffic and new forms of freight handling led to the closure of many railway lines, stations and goods yards. Structural and technological changes in industrial production forced many inner city manufacturing firms to close or relocate elsewhere. The abandonment of over-ambitious urban renewal and urban motorway proposals also created large swathes of vacant land in some areas. Whatever the causes, much of this vacant and derelict land was unsuitable for immediate redevelopment by reason of location, access, size, configuration, pollution or the presence of derelict structures. Dereliction came to characterise parts of the inner city and the pervasive atmosphere of decay became a discouragement to new investment.

In the face of the deteriorating economic situation of the mid-1970s, a series of local policy initiatives were developed in an attempt to stimulate economic growth. These included the establishment of a joint County and District Economic Development Committee, a Merseyside

Economic Forum and a County Economic Development Office.   A 'Merseyside Innovation Centre' was established to support special employment programmes and youth training schemes (Cornfoot T, 1982, p25).

Through the second half of the decade unemployment in Merseyside doubled so that by June 1979 some 82,000 people (1 in 8 of the workforce) were out of work (Merseyside County Council, 1979, p14). Most of the unemployed were men, with a high proportion of being unskilled and/or young. It was claimed that Merseyside had the highest concentration of youth unemployment in Britain at the time. In addition to job losses in the traditional port-related and manufacturing industries, a number of firms that had been encouraged by regional policies to open branch plants in the area a decade earlier were also facing cutbacks or closure. Whilst nationally the decline in manufacturing employment was being offset by growth in the service sector, this was not the case in Liverpool. In part this was because as a port Liverpool already had a sizeable workforce employed in insurance, banking and shipping offices, so that with declining trade, increasing productivity and centralisation of some higher order service functions to Manchester and London, Liverpool lost, rather than gained jobs in many sectors. The County Council concluded that employment on Merseyside would continue to fall in subsequent decades.

During the 1980s the Thatcher government became increasingly frustrated by what they regarded as an inefficient and inadequate local government system that, in their view, inhibited local economic recovery and development. There were a number of high profile political conflicts between the government and inner city local authorities. One of the most vicious of these battles was fought between the government and Liverpool City Council, doing lasting damage to investor confidence and the reputation of the city (see chapter six).   However, it was Merseyside County Council that was abolished by the government in 1986 as an unnecessary layer of local bureaucracy.

It was an irony then that in 1988 a cumbersome melange of 22 organisations including district councils, joint authorities, government agencies and departments were brought together to prepare the 'Merseyside Integrated Development Operation' (MIDO): a bid to the European Community for subsidy to support economic regeneration.   The bid included an analysis of the socio-economic and environmental situation on Merseyside and identified a number of local strengths and weaknesses as the basis for a strategy. These are shown in Table 2.1. The analysis was interesting for its recognition of the local economic importance of tourism and culture and the key role played by the external image of Merseyside in attracting or deterring potential inward investment. In the late 1980s Liverpool was not a city that easily attracted footloose private investment.

Table 2.1   Strengths and weaknesses of the Merseyside economy, 1988

	Strengths	Weaknesses
Land	No shortage of land Buildings available for conversion	Much derelict land in poor condition Historic buildings in need of attention
Infrastructure	Recent investment in water, power and cleaning Mersey	Electricity supply needs upgrading Pollution still a major problem
Communication	Good motorway links Intensive public transport Port returning to profit Airport environmentally well placed but needs investment	Some gaps in the motorway system Modernisation and extension of the electrified rail system needed Port needs more investment and flexible working
Environment	Some excellent urban and rural environments	Much urban dereliction and poor environmental image
Business development	Local market of 4 m people Some large successful local firms, e.g. Littlewoods, Pilkingtons, Unilever	Too dependent upon 'branch plants' Too small a base of small and medium sized enterprises (SME)
Tourism	19 million visitors a year Wealth of proven tourist attractions	Need for pump priming Poor image Environmental neglect
Employment	Recent investment in training	Large numbers of unemployed High proportion of semi-skilled and unskilled in workforce
Image		Poor image of Merseyside is a fundamental constraint

Source: Merseyside Task Force, 1988, Section 7

In 1993 Merseyside was accorded 'Objective 1 status' by the European Union (EU). This status was given to regions within the European Union where the per capita Gross Domestic Product (GDP) was less than 75 per cent of the EU average. Given that Merseyside had not previously been accorded such status this was an indication of the relative failure of economic regeneration policies for the region in the preceding decade. As part of the Objective 1 monitoring process a 'Merseyside Economic Assessment' was prepared in 1995 (Labour Market Strategy Group, 1995). Like the 1988 report, it concluded that Merseyside still lacked competitiveness in comparison with other UK and European regions. Future growth was likely to remain slower than the rest of the North West and the UK as a whole, leading to continuing divergence. Recovery and convergence would only be achieved if Merseyside firms could secure and maintain competitive advantage and if new inward investment could be attracted. Furthermore, in order to maximise competitiveness, those people currently economically excluded (including the unemployed) should be re-integrated into the labour market and society (Labour Market Strategy Group, 1995).

Population, Housing and the Local Environment

Population and households

Between 1951 and 2001 the population of Liverpool fell from 790,838 to 439,476 a decline of over 44 per cent (see Figure 2.4).

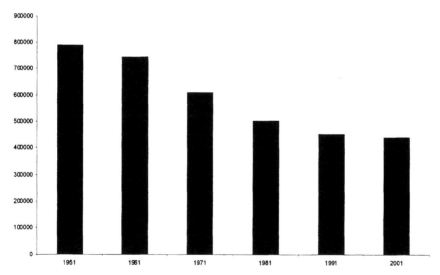

Figure 2.4   Population change in Liverpool

Source: Census of Population

However, the fall in the number of households was less dramatic. Whilst there were 217,594 households in 1961, there were still 182,822 households in 1991, a drop of just under 16 per cent. The reason was that average household size was also falling as part of a national trend. More young and elderly people were living alone or in childless households while many families were having fewer children. In 1961 there was an average of 3.42 persons per household but by 1991 this had fallen to only 2.44. In other words, in 1961 about 292 dwellings were needed to accommodate every 1,000 people in Liverpool but by 1991 the housing stock would have to be expanded by almost a third to 410 dwellings per 1,000 people to house the same population. Thus the number of households fell at a much slower rate than the population.

Over the same period the number of dwellings declined slightly from 208,196 to 206,314 (-0.9 per cent). As a result of these changes the ratio of dwellings to households improved from 0.96 dwellings per household in 1961 to 1.13 dwellings per household in 1991. This is to say, over the three decades the city moved from a crude housing shortage to a crude housing surplus. This change had a number of consequences. With a wider choice of accommodation households were increasingly able to reject the least popular housing (i.e. dwellings in poor locations, overpriced or of poor quality design, construction, amenities or state of repair). This might have been of benefit to consumers but led to housing providers in all tenures being left with surplus stock that they could not sell or let at any price. It also meant that there was an increasing per capita burden of housing maintenance. In 1961 each member of the population, on average, could theoretically be said to be responsible for the maintenance and repair of 0.28 dwellings, by 1991 this per capita burden was about 0.41 dwellings (i.e. each member of the population carried the burden of maintenance and repair of more than a third more dwelling space in 1991 than in 1961). This would have significant financial implications for any city but in Liverpool other features of population change combined to exacerbate the situation.

Between 1961 and 1991 the proportion of the population aged 75 years or over (and therefore likely to need special help and support with housing) almost doubled from 3.4 per cent to 6.6 per cent of the total population. Between 1971 and 1991 the proportion of lone adult families (many of whom had much lower than average household incomes) increased more than four-fold from 1.6 per cent to 7.0 per cent of all households. Similarly, the unemployment rate in Liverpool increased dramatically over the period, although it fell back somewhat during the 1990s (see above). The combined impact of these changes suggests that over the three decades the financial ability of the population of Liverpool to house itself satisfactorily without external subsidy steadily diminished.

The housing stock

One of the biggest changes in the physical character of Liverpool's residential areas has been a steady fall in population density. With a falling population accommodated within the city boundary the overall population density fell from around 63 persons per hectare in 1961 to around 40 persons per hectare in 1991. At the same time there was an increase in the amount of land and buildings within the city used for residential purposes. The density of population within residential areas therefore declined by an even greater amount. Falling densities made the provision of local community services and commercial activity more difficult to achieve. If, for example, it takes an average population of 5,000 to support a two-form entry primary school, then at the 1961 density such a population would have been accommodated on about 0.79km^2 of the city; at the 1991 density the population is spread over 1.25km^2 But between 1961 and 1991 the proportion of children aged 0-15 years within the population declined from 26 per cent to 20 per cent so the average catchment area of a primary school would have further increased to around 1.40km^2. The implication of this was that many children would have to travel further to their primary school, with consequent impacts on the demand for car use, pollution and energy consumption. Commercial activity, such as the provision of local shopping, would also depend upon a local catchment population. Not only did falling densities reduce these local populations in many areas but rising car ownership and the advent of large supermarkets led to a dramatic reduction in local retail provision.

There were also major changes in housing tenure over the period as shown in Figure 2.5. It was a feature of the housing policy of all governments over the period that they strongly supported the growth of owner occupation. Mortgages became easier to obtain and the transfer of existing dwellings out of from council and private renting was encouraged. Since the passing of the Housing Act 1974 housing associations played a growing role in the provision of social rented housing in the city.

The physical composition of the city's dwelling stock also changed with the proportion of detached and semi-detached dwellings rising from 22 per cent of the stock in 1977 to over 26 per cent in 1991 while the proportion of purpose built flats and maisonettes declined over the same period from 20 per cent to around 17 per cent (National Dwellings and Housing Survey, 1978, Census, 1991).

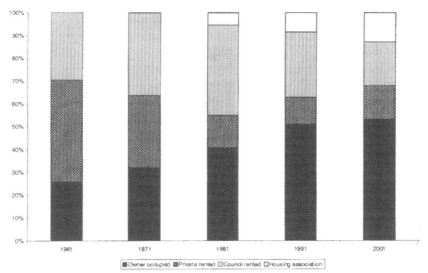

Figure 2.5   Housing tenure in Liverpool

Source: 1961 data: Liverpool City Council, 1987, Past Trends and
                Future Prospects. 1971-91 data: Liverpool City Council, 1993,
                Key Statistics from the 1971, 81, 91 Census. 1997 data,
                Liverpool Housing Needs Survey 1997

These changes reflected the growth in private sector housebuilding and
the demolition of unpopular forms of social housing accommodation, such
as high-rise flats, that occurred during the period. The proportion of
households without exclusive use of all amenities fell from 38.2 per cent in
1961 to only 1.9 per cent in 1991, a dramatic improvement. However, even
in 1991 central heating was available to only 61 per cent of Liverpool
households compared with a national figure of 78 per cent (DCA, 1997,
p14).

The rate of housing completions in the city fluctuated over the period
reflecting economic trends and government priorities but there was a
general trend towards decline in building for the social housing sector and a
rise in the proportion of completions for private consumption (see Figure
2.6).

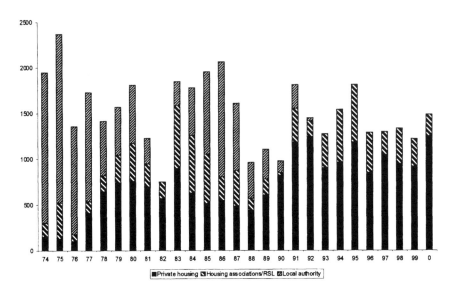

Figure 2.6 **Housing completions in Liverpool, 1974-2000**

Source: Local Housing Statistics HMSO, various years. Until 1979 'housing associations' were included as 'other public sector housing completions'. From 1996 onwards data is for the financial year April-March

By the mid 1970s private speculative housebuilding was contributing less than 10 per cent of all new completions in the city. There were few suburban sites and the risks associated with inner city investments perceived to be too great. This changed in the late seventies with new suburban development within the city boundaries on land released at Croxteth Park and with the emergence of the city council's novel 'build for sale' policy in the inner city. Under this initiative the council encouraged speculative housebuilding in the inner city by selling developers a licence to build private dwellings on council owned land. The effect of the policy was to add around 2,000 private sector dwelling completions between 1977 and the mid 1980s. Although such housebuilding did not directly meet the needs of those on the housing register it did make a contribution to housing supply during a period of very limited council housebuilding. The scheme was also significant in boosting the confidence of private housing developers towards investing in inner city: a crucial factor in supporting many subsequent inner city and city centre housing projects.

The mid 1980s saw a slackening in private housing investment in the city but towards the end of the decade the city was enjoying a modest boom in building for owner occupation. By the 1990s government policy severely limited the scope of local authorities to undertake any substantial housebuilding programme and the city became increasingly reliant on the private sector and housing associations for new housing.

Until the 1970s the contribution of housing associations to overall housing provision in the city was minimal but under the Housing Act 1974 the Housing Corporation was made responsible for the promotion, funding and supervision of housing associations. New grants were made available and local authorities were also encouraged to support the sector. The housing association movement expanded particularly rapidly in Liverpool, increasing their share of the housing stock from around 4 per cent in 1974 to just under 9 per cent in 1991 and rising to over 12 per cent by 1997. The effect of more recent government policies is likely to increase this proportion still further in the near future.

There has been a long run decline in housebuilding by the local authority. This downward trend was well established by mid seventies when the ending of large scale clearance and restrictions on public expenditure caused many local authorities to review their building programmes. During the decade up to 1983 Liverpool City Council was administered by a 'hung council' in which no party achieved overall control, although the Liberals were for much of the time the largest party. This administration struggled to agree a housebuilding programme and local authority housing completions declined sharply. With the return of a majority Labour administration in 1983, there followed a rapid expansion of the local authority housebuilding programme, against the wishes of and without adequate funding from central government (see chapter 6 below). In recent years government policy has effectively eliminated the local authority as a direct provider of new housing, instead preferring to see them in an enabling role supporting and facilitating housing provision by the private sector and other social housing agencies.

The problem of oversupply

Oversupply was one of the key features of the Liverpool housing market to emerge in the nineteen nineties. By this time the number of dwellings exceeded the number of households in the city and the ratio of dwellings to households was more generous than in any other equivalent English city except Manchester. The number of people on the City Council's housing register fell from around 11,000 in 1990 to just over 4,000 at the end of the decade. In an analysis of dwelling vacancy the City Council concluded that there were between 9,500 and 11,000 dwellings across all tenures either long term vacant or surplus above a notional 2 per cent vacancy rate to allow for normal turnover rates (Liverpool City Council, 1998, p8). A lower than average price can also be indicative of oversupply. The average dwelling price in Liverpool in 1996/97 was £53,789, nearly 30 per cent below the UK average figure of £76,623. But there were differences within the city with average prices in North Liverpool (£42,759) being 30 per cent below those found in South Liverpool (£61,411). (DCA, 1997, pp25-26.). Further, the average price of a dwelling in Liverpool in 1997 was 2.74 times average income, whereas the equivalent figure for the UK as a whole was 3.71 times average income (author's calculation).

This was not to say that every household in the city that needed a dwelling could afford one or that further housebuilding was unnecessary. According to the Liverpool Housing Needs Survey:

> The key fact is that, based upon conservative assumptions, access to home ownership is beyond the reach of over half of the (10,400) concealed households identified in the DCA survey on any realistic assessment of availability of properties (DCA, 1997, p28).

And according to the City Council:

> Despite an overall crude housing over supply in the city....there is a need for new building developments in the city to provide the right balance of housing stock in favoured locations and to meet rising aspirations (Liverpool City Council, 1998, p6).

Nevertheless, this housing surplus had a profound effect on the housing market. Although house prices were lower, there was no equivalent reduction in maintenance costs and there was evidence that, particularly at the lower end of the owner occupied sector, householders had some difficulty in keeping up adequate levels of maintenance and repair and a tendency for the condition of this stock to deteriorate (Littlewood and Munro, 1996; Leather and Morrison, 1997). Lower prices also suggested that the relative profit from new building for owner occupation in Liverpool would be lower than the UK average and so discourage new private housebuilding. However, the situation was complex. Clearly in recent years the city had been able to maintain a sizeable rate of private housing completions. It seems that speculative housebuilders were able to divert demand away from the second-hand housing market by offering incentives such a cheap mortgages and payment of legal fees. Furthermore, many schemes in the city, particularly in the inner city and city centre benefited from subsidies, such as those on derelict land reclamation, or partnership arrangements with the social housing sector that reduced financial risk. In consequence this housing surplus, rather than discouraging new housing investment, appears to have resulted in reduced investment in the existing stock as well as a rising vacancy rate, particularly amongst dwellings that were unpopular by virtue of location, tenure, price or amenities.

Desipte this surplus some groups still found it difficult to obtain suitable housing. During 1997/98 the council housed some 840 homeless households in priority need and identified a further 600 cases of homelessness but not in priority need (Liverpool City Council, 1998, p9). There was also a shortage of suitable housing for those with special needs, notably the elderly and disabled. Two reports on housing and race in Liverpool expressed concern about racism within the Liverpool housing system (Commission for Racial Equality, 1984 and 1989). In 1998 the City Council noted that despite a pattern of income distribution similar to that of white communities the black and ethnic minority population continued to experience

> low levels of owner occupation, with substantial numbers of households being housing association or Council tenants....a significantly higher level of dissatisfaction with present accommodation.....a marked tendency for less stability, measured in terms of length of stay at their present address, than their white counterparts....and significant experience of racial harassment (Liverpool City Council, 1998, p10).

The residential environment

The residential environment also changed with long term improvements in many parts of the city. It is easy to forget the impact of the 1956 Clean Air Act, which together with the installation of new heating systems through housing renovation, virtually eliminated the sulphurous smoking chimneys that typified and polluted the winter streets of the inner city until the sixties. The decline of local industry and changing technologies of production similarly contributed to a cleansing of the residential air and a reduction in the intrusion of industrial traffic. On the other hand the rising car ownership brought increased danger and pollution from through and parking traffic in residential areas.

The greatest changes were seen in the inner and outer council estates and in the inner city private housing areas. By the late 1970s local environmental conditions in many of the city's council estates were at an all time low: abandoned and derelict properties, ill-maintained landscape, vandalised facilities, glass-strewn roads and parking lots and graffiti covered walls were an all too common image. Subsequently, a series of powerful spending programmes, ranging from the early 'intensive management schemes', through the 'urban renewal strategy' of the mid 1980s, to the 'Estate Action' programmes of the 1990s, all had a dramatic effect in improving the local physical environment. The townscape of the council estates was further changed by the demolition of many tower blocks under the policies of the Liverpool Housing Action Trust (LHAT). Similarly the inclusion of environmental works such as traffic calming and landscaping within the city's 118 Housing Action Areas and General Improvement Areas during the 1970s and 1980s brought about less spectacular but nevertheless beneficial improvements to the environment of the older private housing neighbourhoods.

Summary

The port and port related industries were central to the growth and development of Liverpool over many years. However, since the middle of the 20th century their importance in the local economy has gradually declined. Whereas the sector was once the dominant employer in the city it has today become insignificant. Related industries have become footloose, driven by economies of scale and integration to locate where they will.

Other industries too have moved out of the inner city. Some have closed or merged with competitors in other locations, some have decamped to the periphery of the conurbation in search of cheaper land and better access to the inter-regional motorway network. New industries have been attracted to the conurbation but the majority of these have tended to locate on at the edge of the conurbation in Halewood, Runcorn or Ellesmere Port. Very few manufacturing firms have been attracted back to the inner city.

Commercial investment has been attracted and some of this has been clustered in the city centre but much of this too has been disapated across other parts of the conurbation. It is in this sector particularly that the city has lost out to competing centres. Chester, Warrington, Manchester and other areas have out-competed Liverpool for retail sales and investment. Manchester, with its international airport, better regional accessibility and more dynamic local economy has become the dominant regional centre, attracting far more office investment than Liverpool over the last three decades. Suburban office parks such as Daresbury, near Warrington and Chester Business Park have attracted 'back office' employment. Even in the growing economic sectors of culture and tourism Liverpool is struggling to compete with Manchester. Liverpool has some of the finest museums and galleries outside London and the excellent Philharmonic Hall but recently Manchester has seen investment in similar facilities together with regional exhibition space (GMEX), major conference and sporting facilities. All of these investments together create an upward spiral of growth in the Manchester economy that is less evident in the Liverpool economy.

Image is said to be one of the problems detering investment in Liverpool. This seems to have intensified as the recession began to bite in the early seventies and again during the period in the mid-eighties when the 'militant' Labour Group were in control of the City Council. Subsequently, there have been some improvements but image and perception still seem to be significant deterents to inward investment. Nevertheless, Merseyside still remains a very large local economy employing over half a million people producing a Gross Domestic Product exceeding £1.5 billion a year (Labour Market Strategy Group,1995).

Associated with these economic changes has been a fall in population. A proportion of the population loss has been caused by out-migration, especially by young people leaving the conurbation in search of better economic prospects elsewhere. Others have moved from the inner city to the periphery of the conurbation either driven by social housing policy (mainly before the mid-seventies) or as growing affluence has allowed the purchase of homes in more attractive suburban locations. There has been a tendency for the old and the poor to be left behind in this process. Household size has continued to fall and the rate of natural increase amongst the residual population is low. These factors have reinforced a downward spiral of population decline that is hard to break.

While population and housing densities have fallen, so housing conditions have improved and tenure has changed with over half of all households now living in owner occupied accommodation. Until the early seventies slum clearance was the main policy for dealing with obsolete housing but for the next twenty years an alternative policy of housing renovation was vigorously pursued. However, by the millennium there were signs that this approach has run its course and other approaches would be needed in future to maintain housing quality and keep demand and supply in balance. One of the biggest changes in housing investment has been the return of private housing developers to the inner city. One of the most interesting recent trends has been the emergence of the city centre as a popular residential location, especially amongst young households.

The provision of shopping and other services has become increasingly centralised into fewer, larger units as falling densities and rising car ownership have combined with retailers' and service providers' needs for ever greater economies of scale. The result has been the closure of many local shops, social, health, welfare, recreation and leisure facilities.

By and large the physical environment of the city has improved over the period. Virtually all housing is now provided with full amenities although too much of the older stock is in an inadequate state of repair. Much of the dereliction that appeared in the seventies has since been treated and the land re-used. The townscape has changed: much of the high density traditional fabric of the inner city was swept away in the fifties and sixties to be replaced by multi-storey blocks of flats. Today many of these same flats are being removed to be replaced by low rise, low density housing. Many traditional shopping streets have declined in importance as large supermarkets and retail parks have been developed in different parts of the city. The City Council has successfully protected numerous conservation areas from inappropriate development. Much of the city centre has been pedestrianised and the traffic has become much better managed and controlled in recent years.

Compared with other cities in the UK and even across western Europe, the city's economic performance has deteriorated since the early 1960s. Economically the city is much less important than was the case then. This has had consequences for social and environmental conditions. Public agencies have struggled to cope with high levels of social exclusion, poor health and crime. At the same time, for most inhabitants, there have been considerable improvements in physical conditions and the local environment. Indeed, for those in stable and reasonably paid employment, the city continues to offer a high quality of life.

3 Modernising the City

According to Taylor (1998) town planning in post-war Britain could be distinguished by three related features:

- concern for the *physical* structure and form of towns, that is to say it paid relatively little attention to the economic or social analysis of towns or to socio-economic policy
- emphasis on a *design* process, essentially based on artistic rather than scientific principles
- concern with the production of *'end-state' or 'blueprint' plans* that would indicate some ideal state which would be reached in the physical development of the town, i.e. a town plan could be compared in character, if not in scale, to the plans produced by architects and engineers for the production of artefacts

It therefore followed that planning was in large measure seen as a *technical* process, applying the professional skills of the planner, rather than a political process of mediating between the competing demands of various problems and groups in society (Taylor, 1998, pp5–17). However, by the 1960s planners had begun to view the town not as a static entity but as something dynamic and constantly evolving. The town came to be viewed as a 'system' of interconnected parts. Some years earlier two American researchers (Mitchell and Rapkin, 1953) had established that traffic generation was a function of land use. That is to say different land uses generated and attracted different amounts and types of traffic. Thus the disposition of land uses had a direct influence on traffic and an indirect influence on congestion. If the town was viewed as a system, then it followed that the system had to be understood through scientific investigation, and the consequences of actions explored, before rational decisions could be made about land use zoning or transportation investment.

Thus planning had to evolve from being an 'art' concerned with designing towns to encompass the 'science' of planning. Thus it became possible to think of much of town planning as an activity that required an analysis of the whole town or city on a scientific basis in order to produce a rational plan for the efficient organisation of the 'urban system'. Nevertheless, there remained a more local scale at which it was still possible to think of planning as 'urban design'. There was still scope for the 'design' of places (streets, squares, parks) larger than individual buildings but smaller than towns or districts and the preservation of built heritage, according to artistic rather than scientific principles and methods. This was one of the theoretical arguments in favour of a new two-tier approach to town planning that separated strategic urban policy making from local planning and design.

There were other, more procedural arguments for such a split. Under the Town and Country Planning Act 1947 all Development Plans produced by local planning authorities had to be submitted to the Ministry of Housing and Local Government for approval. This involved the Ministry in decisions about local land allocations and design. It was argued that this was unnecessary and that central government needed only to be concerned with general policy and major proposals. Thus the whole plan making process might be speeded up and made more efficient if the strategic level of planning, in which the Ministry had a legitimate national interest, could be separated from decisions of purely local significance.

Furthermore, if the town was to be perceived as a dynamic system, it followed that urban change and evolution would be perpetual and that the concept of the 'end-state' was redundant. Except at the most local scale, in could no longer be thought appropriate for planners to mimic architectural practice in the production of 'blue-print' plans. What was now required was an ongoing planning *process* that monitored the urban system and intervened to rectify faults or to create improvements.

Responding to these concerns the Ministry established a 'Planning Advisory Group' to consider changes to the planning system. Their report, published in 1965 (MHLG, 1965) proposed that a new type of development plan should comprise two distinct types of plan: a strategic or structure plan would be prepared for a whole urban area or county and be subject to approval by the Ministry. These plans would be concerned with general policy and major proposals and be subject to a continuous process of monitoring and review. Within this strategic framework a lower tier of tactical or local plans, concerned with physical planning and design, would be produced and approved by local planning authorities as and when necessary.

In the early 1960s Liverpool was one of a number of British cities at the forefront of innovation in planning. A 'city planning' department had been established in 1961 under the direction of Walter Bor. The consultants Shankland, Cox and Associates had been appointed to provide additional professional advice. In 1962 the Merseyside Conurbation Traffic Study was one of the earliest in the country to cover the wider field of transportation including public transport. From early in the decade the city was widely recognised for the vigour of its slum clearance programme and its embrace of industrialised methods in council housebuilding. Liverpool was an important city facing major urban problems and had a capacity for town planning that was second to none in the country. It is therefore not surprising that the two plans produced by Liverpool City Council (the city-wide Interim Planning Policy Statement and the City Centre Plan) were amongst the earliest in Britain to embody these new theoretical and procedural approaches to the planning process.

Liverpool Interim Planning Policy Statement, 1965

The Liverpool Interim Planning Policy Statement (IPPS) (1965) was by no means the first plan for Liverpool but it did represent one of the earliest attempts to prepare a city plan that went beyond the land use zoning function of earlier statutory Development Plans prepared under the Town and Country Planning Act, 1947. This plan offered a vision for the future of the city as an interconnected system with economic, social and physical dimensions. Prepared after the publication of the Buchanan Report (MHLG, 1964) and strongly influenced by the debates that were taking place within the Planning Advisory Group, the IPPS was in many ways the first prototype structure plan in the UK.

The title itself is illustrative of a new approach: it was a planning *policy statement*, not merely a land use plan but something more – a set of policies about the future of the city. Nor was it just a drawn plan – it was essentially a written statement containing both a reasoned analysis of the present city structure and problems and a set of policies to solve these problems within a new city structure. This is not to say that many previous plans had not been based upon a written analysis and policy statement – we only have to look at Abercrombie's Greater London Development Plan (1944) or more locally at Longstreth-Thompson's Mersey Regional Plan (1945) – but it did represent a more sophisticated approach than that which had been adopted by most local planning authorities during the previous two decades. It was also a radical plan insomuch as it proposed the comprehensive redevelopment of much of the city centre and inner areas together with major investments in both road and rail transport infrastructure.

The document had the appearance of a high-quality book rather than a local government report. It was liberally illustrated with maps, plans, sketches, diagrams and photographs: clearly designed to be accessible and easy to read. The forward, written by Alderman W. H. Sefton, Leader of the City Council, gave an indication of the new awareness being shown by the plan:

> Even in planning itself we face new situations every day - no longer can we plan in three dimensions only, and our planners must be continually aware of the ever-changing requirements of our people in culture and recreation, amongst other things. Economic considerations force themselves into the picture and pertinent questions cry out for answers. Can we afford to allow penetration of our City by the motor car to such an extent that life becomes unbearable? How much motorway construction can go on? Is our City, like some in other parts of the world, to become a jungle of concrete pillars linked by concrete ribbons, strangling the City in one great gargantuan knot? Or are we to find the necessary amount of money to provide a proper public transport system? These are only some of the questions that must not only be faced and answered by technical officers, but must be weighed by us all, elected members of the Council and citizens. The answers must be found as a result of co-operation by all people who love Liverpool as a place where a good life can be lived. A City for people can only be planned and designed if there is a two-way traffic of ideas between planners and people, and it is in that spirit that I commend this report to its readers. I sincerely hope it is widely read and debated so that when the new Liverpool begins to take shape we can all justly and with pride look upon what we have created and say THIS IS LIVERPOOL, OUR CITY (Liverpool City Council, 1965, p9).

The statement revealed a great deal about the City Council's view of planning at that time. It showed awareness that planning needed to take account of the economic and social problems as well of purely physical issues. There was also a recognition, no doubt influenced by the Buchanan Report, that increased motor vehicle usage posed difficult questions for cities about traffic management and control. The importance of investing in public transport was understood. Alderman Sefton's comments also revealed some interesting views about the planning process. There was an acknowledgement of the professional role of 'technical officers' in supplying answers to these questions and, perhaps for the first time in a plan, recognition of the importance of public participation in the debate about future policy and the idea that the results of the plan should be 'owned' by all the people of Liverpool, not just the City Council.

The city analysed

The first half of the document analysed the existing city structure and its problems in six chapters: the city today; the regional setting; transport; the centre; working areas (sub-divided into the dockside belt, middle city industrial areas and outer industrial estates); and living areas (sub-divided into the inner, middle and outer residential areas) (See Figure 3.1). Thus, the analysis was partly by function and partly by spatial area. It was the spatial analysis that tended to dominate subsequent chapters and this allowed the plan to discuss the inter-relationships between land uses in a way that a function-based structure would not have permitted.

The origins and historical development of the city were briefly outlined. The city structure was contrasted with that of Birmingham and Manchester and shown to be significantly more fragmented than either. The importance of the River Mersey, which divides the city from much of its urban hinterland was seen as a key structural feature. Further influences on the evolution of the city's structure were identified as:

- the city's site and topography
- the radial growth of its transport system
- the dominance of its central core
- forces making for inner concentration (centralising tendencies)
- forces making for outward growth (decentralisation, suburbanisation)

Liverpool was then the nation's second port and important to the economy and social life of a region extending from North Wales to the Lake District and from the Irish Sea to the Pennines. The problems of the city region (essentially the area bounded by and including Southport, Ormskirk, St Helens, Warrington, Chester and Deeside) were said to be characterised by 'over-concentration of services' and a still insufficient provision of 'growth' manufacturing industries. This imbalance was seen as a legacy of 19th century economic growth and structure. Despite the provision of new industrial estates and the arrival of the motor vehicle industry the report notes that:

> this imbalance contributes to the problem of unemployment, migration from the region, and a relatively unprosperous community (Liverpool City Council, 1965, p24).

Unemployment was seen as a male problem (female unemployment was reported as 1.8 per cent, male unemployment 4.6 per cent). At the same time there was a shortage of skilled workers in certain economic sectors and a call for a 'training programme fully geared to the needs of Merseyside'. It was also acknowledged that poorly planned urbanisation had led to a confusion of traffic movement and land uses resulting in an uneconomic and wasteful use of land. The separation of residential areas from major employment zones had resulted in long journeys to work that contributed to traffic congestion, particularly in the city centre. The analysis of the transport system was comprehensive and included the road network, bus services, rail transport, ferry services, the docks and air transport. Mobility, incomes and leisure time were all on the increase in 1965. Two major trends are seen as occurring simultaneously: transfers from rail to road transport and from public to private transport. The road network was perceived to be congested, lacking adequate capacity and suffering deteriorating environmental conditions. The absence of major north-south cross-city routes and reliance on only one cross-river route were identified as strategic gaps in the system. The speed and reliability of bus services were also adversely affected by these traffic conditions. Further problems included through traffic filtering through residential areas and pedestrian safety in the vicinity of main roads. In transport, the underlying conflict between short-term financial objectives and longer-term planning social and environmental aims that prevades all city planning was evident and well illustrated by the following paragraph on the role of the local railway system:

> An attractive rail passenger network as an integral part of the conurbation transport system would play an essential role in alleviating overloading of the road network. The British Railways Board, however, has proposed the closure of nearly all the local passenger services, on financial grounds. It is obviously necessary to consider the cost of all transport services, including social costs, when planning for a satisfactory combination of different modes of transport (Liverpool City Council, 1965, p28).

This statement showed the City Council's awareness of the role that railways could play in an integrated local transport system but highlights the failure of the government to devise and implement a system of accounting that took adequate account of social costs and benefits. The IPPS concluded that an efficient and attractive public transport system was the only means of reducing demand for road travel; that environmental and traffic management techniques must be used to improve life in the city and channel vehicular traffic onto routes which could best accommodate it; and that full use would have to be made of existing roads and railways within the limits of good environment and efficient land use. To this extent the plan demonstrates an awareness of green economics that was ahead of its time.

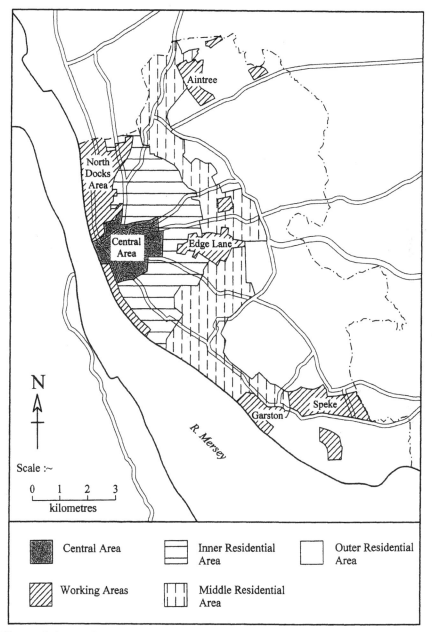

**Figure 3.1   Interim Planning Policy Statement 1965: the analysis of
Liverpool's planning problems by zone**

Source: Liverpool City Council, 1965

According to the IPPS the city centre was a focus for all Merseyside, providing shopping and wholesale distribution, entertainment, cultural facilities and employment for a hinterland of over two million people. One in eight Merseyside jobs were located within Liverpool city centre and it accounted for 26 per cent of the conurbation's retail trade, compared with 14 per cent in Manchester and 13 per cent in Birmingham. Despite problems of obsolescence and congestion there was no anticipation that anything other than expansion lay ahead for city centre shopping and commercial activity. To overcome these problems and facilitate growth the city centre was to be subject ambitious proposals for restructuring and rebuilding.

> The high intensity of land use has added to the difficulties associated with renewal and expansion in the centre. Some 60 per cent of the functions of the area are accommodated in buildings unsuited to their purpose. On the fringe of the city centre numerous small scale businesses that occupy low rental property are likely to be disturbed or displaced, either by pressure for the expansion of true central area functions or in the course of general redevelopment, and will therefore present an important relocation problem (Liverpool City Council, 1965, p32).

The tone of the statement was modernist, expressing a desire to increase efficiency of individual buildings and to sweep away 'obsolete' premises. There was no conception at this time that a range of old and new, high and low value properties might be essential to the efficiency of the city centre as a whole. Nor was there any notion that these older buildings might have a value other than that which can be measured in rent: for example their value as heritage or as landmarks for local people.

The main problem of the 'working areas' of the city was seen as a shortage of land for new industrial and service development to meet the growing employment needs of the population. It was anticipated that this need would be exacerbated by the effects of the urban renewal programme as it increased the demand for land by displacing many businesses in the older areas. It was recognised that due to shortage of land within the city, most of the growth of employment, especially in manufacturing, must occur in neighbouring authorities beyond the Liverpool boundary. Despite a recognition of higher than average unemployment and a relative lack of prosperity there was no anticipation of the industrial collapse and restructuring that was to hit the city less than ten years later. In the mid-sixties the economy of the city seemed fairly secure. It was still an important port and a major city with a substantial industrial base set within a national economy that had been growing steadily for more than fifteen years.

The city's living areas comprised not only dwellings but also shops, schools, open spaces and social facilities. According to the plan, some of the city's most serious planning problems were to be found in these areas, including obsolete buildings and layouts, inadequate social facilities, and an environment eroded by the intrusion of motor vehicles. The plan recognised that changing patterns of social and economic demand would accelerate the obsolescence of some types of existing development; and that there would be increases in incomes, leisure times and car ownership together with changing patterns of consumer spending. A number of population trends were also identified:

- continued decline in the total city population
- redistribution of population from the inner to outer areas
- an unbalanced pattern of out-migration reducing the proportion of children and younger people in the residual population
- declining household size.

Sixty per cent of the city's dwellings had been built before 1920 and some 79,000 (over 38 per cent of the total stock) were identified as unfit. Rising consumer spending was not being reflected in increased shopping floorspace as retailers became more efficient and supermarkets and multiple stores began to oust small local traders from their traditional role. This was an early indication of more radical developments that were to emerge in subsequent decades. Combined with the outward migration of population these trends had the effect of drastically reducing the need for shopping floorspace in the inner areas. The most serious educational problems were seen as the high proportion of obsolete school buildings, sub-standard sites and a shortage of school playing fields. A linked problem was the shortage of public open space within the older districts.

The living areas were divided into three concentric rings. The inner residential area comprised those districts surrounding the city centre, many of which had been or were proposed to be demolished and redeveloped under the urban renewal programme. It was in this area that the worst pedestrian-vehicular conflicts were identified. Here also the overprovision of shops and community facilities and underprovision of good quality school buildings, playing fields and open spaces was at its worst. The physical layout of these older areas was perceived as congested and the townscape was said to be dreary, consisting of 'a monotonous uniformity of terraced houses lining streets arranged in rectangular grids', worsened by pockets of dereliction and the intrusion of 'alien' uses. Commenting on the areas that had recently been redeveloped an interesting bit of self-criticism emerged.

> The use of tall slab blocks in redevelopment, particularly on the slopes of the (Everton) ridge, has extended the 'city scale' outwards from the Centre into this Area. Unfortunately, however, the siting of these blocks has marred the shape of the ridge, one of the few topographical features which allows the observer to locate himself within Merseyside (City of Liverpool, 1965, p49).

The subsequent removal of many of these same blocks in the 1990s has restored much of the view of Everton ridge.

The middle residential area comprised some rather better (later) pre-1920s housing, together with some former village centres. It therefore had more variety of townscape and character and was generally in better condition with fewer problems than the inner residential areas. It was here that some of the city's biggest parks were to be found, such as Stanley Park, Newsham Park, Princes Park and Sefton Park. The intrusion of industry, both large-scale installations as well as local non-conforming uses, was seen as problematic. Most of the housing in this area was expected to last a further 15–35 years but it was recognised that they could degenerate into slums if neglected, so action to arrest the process of obsolescence was considered important. The Outer Residential Area comprised that part of the city (about half its total land area) that had been developed since the end of the First World War. It was an area where few problems were identified by the report.

The whole analysis of the city's structure and problems ended with a synthesis of 'the total problem' in which it was concluded that

> The basic aims of the plan must be a high standard of environment within a well-organised framework for economic and social development. The plan must hold good both in the short-term and the long-term, maintain the heritage of the city and promote creative design in new development (City of Liverpool, 1965, p57).

The plan

The second half of the document on the 'Future City Structure' was introduced by a chapter entitled 'A New Type of Plan'. In line with the thinking behind the PAG Report (MHLG, 1965) it was argued that the existing style of Development Plan was ineffective in dealing with either long-term strategic problems or with the more immediate problems of three-dimensional urban design. A new type of plan was therefore necessary: one that would combine clarity and firmness of principles and policies, with flexibility of implementation. A strategy was needed that would govern the overall nature and direction of future *growth* without tying down every piece of land to precise long-term land use allocations. For Liverpool, such a plan would be based primarily upon a comprehensive transport network, closely related to the principal traffic generating uses (Liverpool City Council, 1965, p62). In other words, a strategic land-use transportation strategy, to be followed by more specific plans for individual districts and detailed three-dimensional plans for action areas. Such an approach, it was argued, would enable planning resources to be concentrated on problems as they arose, rather than spread thinly over the whole city. What followed in the document was such a strategy. It was 'interim' only because the City Council had recently commissioned but still awaited results from the ambitious Merseyside Area Land Use/ Transportation Study (MALTS).

The key projection supporting the strategy was that the population of the conurbation would rise substantially over the coming decades. The logical consequence of which would be a need to provide more land for housing and employment. A marked expansion of port activity was even predicted. Unfortunately, as we now know, this rate of growth continued only for a short time after 1965. By the early seventies growth had turned to decline and much of the basis of the strategy was undermined.

Proposals were to be set within the regional context and sought to achieve a number of objectives including:

- an efficient and integrated city and regional transport system, closely related to land-use
- an urban form that would concentrate activity, such as employment, business, shopping and other facilities, in a balanced and decentralised pattern, and closely related to the transport system
- an increase in the amount and range of employment in the city
- greater quantity and choice of housing to meet growing housing needs, together with necessary educational, social and open space provisions
- arresting the decline of districts which would not be redeveloped for some time through a concerted policy of housing and environmental rehabilitation

- re-planning of the whole city on the basis of environmental areas, having a high standard of layout, architecture, landscaping and no extraneous traffic.

Three alternative transport systems were considered. System A attempted to attract the majority of trips onto rapid rail services leaving the road network to accommodate essential traffic plus a relatively low level of other traffic. System B sought to attract the majority of trips onto rapid bus services, while System C tried to cater for a high level of private vehicle movement with only a low level of public transport provision. Both in terms of effectiveness, cost and speed of implementation system A (rail) was seen to have substantial advantages over the alternatives. Thus the planners reached the conclusion that the solution to the long-term transport problem was to provide a greatly improved, rail-based, public transport system and a primary road network which would cater for only a limited amount of traffic beyond essential users. Specific proposals included a city centre underground rail loop to connect the three main terminal stations and a link line to connect the northern and southern suburban railway systems. The upgrading of an existing freight line to form an outer rail loop for suburban passenger services was also proposed.

Decentralisation of the urban structure to relieve congestion pressures in the centre and a close relationship between land use intensity and transport provision were seen as essential.

> The existing radial pattern of major lines of communication and urban areas in Merseyside lends itself to controlled linear development, as a means of ordered regional expansion and decentralisation. Increased efficiency of these 'fingers', by relating high density development to rail and road facilities, offers a great opportunity to exploit the advantages of this radial form, and, in particular, the carrying potential of public transport (City of Liverpool, 1965, p76).

It was also argued that decentralisation would tend to promote suburban centres. Thus district centres would be suburban concentrations of shopping, social, cultural and local employment facilities, and would become major features of the future city structure.

Whilst it was not seen to be the role of this document to develop detailed plans for every part of the city, it was thought necessary to set out the principles by which individual districts would be planned: to establish the ground rules for local urban form. In this context transport was again seen as one of the key determining features of the physical plan. The whole city was to be divided into a series of 'environmental areas' defined by the primary and secondary road layout. Within these areas, although they might differ in size and function, there would be a freedom from through traffic and improvements to the quality and convenience of the local environment. Non-conforming industrial uses were to be relocated but where possible re-sited within the same general area so as not to unduly disrupt the pattern of employment opportunities and the interdependence of related commercial and industrial activities. In slum clearance areas redevelopment was to be comprehensive and carried out quickly. This was because it was thought that only in this way could the Council ensure proper provision for the motor vehicle within an integrated layout for housing, schools, green spaces, shopping, social and other facilities (City of Liverpool, 1965, p85). Local shops, primary schools, childrens' play areas and other local facilities were to be provided within a safe walking distance of every home. Outside the slum clearance areas, older housing was to be rehabilitated and local environmental improvements undertaken. Environmental management would secure better and safer use of the existing road system, the removal of 'bad neighbour' land uses remedying deficiencies in social and community facilities. A number of 'district centres' were to be created around the city, some through the comprehensive redevelopment of existing 'high streets', others would be entirely new (e.g. Belle Vale). Deficiencies in public open space were to be remedied throughout the city with the provision of a major park of at least 10 acres (4 hectares) within about ¼ mile (400m) of every home. Obsolete schools would be replaced on sites provided, wherever possible, in accordance with the minimum standards laid down by the Department of Education and Science. Through such policies the physical urban form of the city was to be *modernised*.

The value of a good urban environment was recognised, not only as a means to attract new business and improve its economy, but also to retain workers who might otherwise join the 'drift to the south' in search of better conditions (City of Liverpool, 1965, p109). The specific character of the historic townscape and the city's heritage was also identified as a matter of importance.

> The bold, distinctive and varied character of the townscape in many parts of the city, the rich collection of fine buildings, the beauty of its parks, are all part of this heritage. Unsympathetic treatment of these buildings, and erosion of some of this existing fine environment by indiscriminate routing of traffic, could do irreparable damage and seriously diminish the stature of the city.....equally important is to ensure that the form of any new development is carefully integrated with the old, so that, for example, the scale, colour and materials are sympathetic, important vistas are maintained, and that the

generation of additional traffic will not destroy the quality of the local environment (City of Liverpool, 1965, p110 and p112).

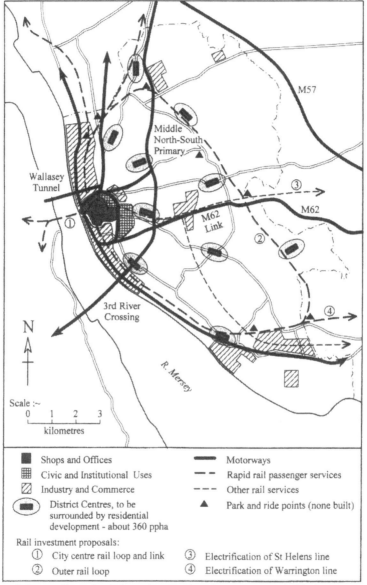

Figure 3.2   Interim Planning Policy Statement 1965: future city structure

Source: Liverpool City Council, 1965

Thus the IPPS represented a new and radical approach to city planning. Although providing a comprehensive strategy it made no attempt to offer detailed solutions for each area. That was to be left to a 'mixed-scanning' approach whereby the resources for detailed plan making were only brought to bear on problems as they were perceived, leaving the rest of the city to the 'light touch' of strategic principles and general policies. This separation of strategic policy from the tactics of implementation was in line with the emerging recommendations of the PAG. In this sense the plan was a prototype for many urban structure plans that were to follow in later years in other cities.

It is notable that traffic was regarded as the dominant problem facing the city. Almost every chapter devoted some space to recording the damage caused by excess and uncontrolled road traffic or proposing some part of the solution. The plan followed very much the philosophy and analysis of the Buchanan Report but appeared to go much further in its support for investment in local rail-based public transport infrastructure. By referring to the intellectual poverty of the Beeching Report (the logic underpinning the British Railways Board's rail closure programme) the plan neatly drew attention to the conflict between long term and short-term public policy objectives and the failure of government of the day to take adequate account of social costs and benefits in its decision-making.

The philosophy on which the plan was based could be described as modernist. That is to say, firstly, the city was perceived in predominantly physical terms. The analysis of spatial structure into concentric rings and sectors relied heavily on the theoretical approach of the 'Chicago School' of geographers. Secondly, the plan sought the best use of technological and organisational innovation to create a city that was efficient, healthy and safe. To put it another way, the principle mechanism for problem solving was the application of technology with little regard for cost and little consideration of alternative approaches. Thirdly, relatively little concern was shown for the retention of older, less efficient, structures or districts when compared with later plans. There was little recognition given to the merits of environmental familiarity, neighbourliness, the social or cultural life of the city. Although there were hints of an understanding of the emerging literature on townscape (see, for example: Nairn, 1955; Cullen, 1961), there was little to suggest that the authors of the plan were responsive to the work of Jane Jacobs (1961) and her analysis of the value of traditional streets, mixed communities and neighbourhoods; or the sociological works of commentators such as Young and Willmott (1957).

The biggest problem with the plan did not emerge for a number of years: that was the assumption of continuing growth upon which the whole edifice of city renewal had been based. By the time of the 1971 Census it had become clear that the population of the conurbation was no longer growing and the city itself was in decline as the effects of slum clearance and overspill began to bite hard on the inner areas. A couple of years later recession had set in and the consequent economic restructuring undermined many aspects of the city's once powerful local economic base. By the mid-seventies there was little money left to implement such expensive programmes of city renewal.

Implementation

In its implementation some parts of the plan fared better than others. The Merseyside Area Land Use / Transportation Study (MALTS) was started in 1966 and completed in 1969. Traffic was seen as an issue that required co-ordinated study and policy across the whole conurbation and MALTS was commissioned by a consortium of all the adjoining local authorities. This study was more sophisticated than previous traffic surveys and plans for the area.   Mobility was considered as a problem for the first time on Merseyside. The performance of different land use patterns was measured against traffic generation and alternative transport systems were tested for efficiency, effectiveness and environmental impact.   Indeed, MALTS provided one of the earliest land use-transportation studies in the UK and became one of the first local authority plans to recognise problems of mobility and to use goal achievement techniques in its analysis. MALTS provided a Merseyside-wide framework for transport planning and confirmed the value of the proposals contained in the IPPS.

Construction of the city centre loop and link underground railway line commenced at the end of the decade and was completed by the mid-1970s. Even as the project was being finished the effects of the 1973/4 recession were leading to public spending cuts and non-essential elements of the scheme were abandoned.   For example, some escalators were never installed and were replaced by stairs. Had the project started only one or two years later it might never have been completed. As it was Liverpool got its underground loop and link lines by the skin of its teeth.   Other projects that were to be implemented later in the programme, such as the outer rail loop and a many of the park and ride interchanges, have never been completed.

Little of the primary road network has been built. Such massive urban road systems involved hugely expensive programmes of land acquisition and demolition as well as environmental damage to adjoining neighbourhoods. By the early 1970s there was mounting opposition to the environmental and social impacts of urban motorways. After 1974 with growing pressures on public spending there was little political appetite to continue such projects. Locally, the new Merseyside County Council undertook a review of highway projects and abandoned virtually all the schemes within Liverpool, with the notable exception of the inner ring road, which was downgraded to a surface-level highway.

Despite the theoretical advantages of the 'finger plan': an urban structure based on decentralised concentration (a concept that still finds favour with environmentalists thirty years later), little of it was actually achieved. Firstly, most urban expansion was to take place outside the city, beyond the control of the Liverpool city planners. Secondly, efficiency in transportation was only one of many competing factors determining the rate and location of urbanisation, the use of land and the density of development. There were many other influences, for example land ownership, housing market pressures and competing land uses. Perhaps it was over-ambitious to expect the political system to place the aim of transport efficiency above other considerations.

Nevertheless, the decongestion of the inner areas did take place. Indeed, the rate and nature of depopulation became one of the major problems of the inner city during the next decade. Overspill took former inner city residents to peripheral council estates in Kirkby, Cantril Farm and Netherley, and to the new towns of Runcorn and Skelmersdale. Even in these planned developments, location and internal layout had little regard to the finger-plan concept of a rail-based development corridor, or to the idea of decentralised concentration. The ambitious housing renewal programme was implemented with great vigour although, as Muchnick pointed out at the time, the comprehensiveness and co-ordination that was so much a part of the plan, was far from being the reality of implementation.

> although the planning department conceives of renewal in comprehensive terms, although its district plans propose the replacement of facilities other than housing, and although the city council reasserts its intention to pursue such a goal, the corporation's policy is in fact a housing programme coupled with hopes and prayers for ancillary development (Muchnick, 1970, p80, quoted in Gibson and Langstaff, 1983).

But this seems to have been a harsh criticism of renewal policy. There was some success in creating 'environmental areas' through the use of traffic management measures to keep through traffic out of residential areas. Slum clearance did effectively remove bad-neighbour industrial uses from residential areas although it was later to be criticised for producing mono-functional districts containing few local employment opportunities. The City Council also invested heavily in the replacement of obsolete schools and other social facilities. However, the policy of creating district centres became problematic and is discussed later.

Although even today much of the city's employment is located in the zones identified in the IPPS, the economic or 'working area' proposals contained in the plan were geared to the steady growth of the 1960s and were completely overwhelmed and rendered near-irrelevant by the recession of 1974. The protection of heritage, which got little more than a tentative mention towards the end of the IPPS, has grown to become one of the strongest elements of planning policy and of significant economic benefit in city marketing and the promotion of tourism.

Liverpool City Centre Plan, 1965

Only months after the approval of the IPPS, the Council published its Liverpool City Centre Plan (City Centre Planning Group, 1965). This plan had regard to the city-wide strategy but was produced in parallel by a joint inter-professional team drawn from the city planning department and their consultants: Shankland Cox and Associates. Three years earlier, in the wake of growing concerns about emerging conflicts between development pressures, traffic and amenity, the Ministry of Housing and Local Government had begun to encourage local authorities to produce new plans for their town and city centres (MHLG, 1962). Liverpool itself had begun such planning a year earlier even than that.

In his forward, Alderman Sefton, Leader of the Council, again set out his views on the purpose of the plan.

> It is to provide an even better city in which to live, to work and to play. Nor is it to be a city only for its citizens; it is to be a city in which visitors will find visual pleasure, colour, excitement, activity and recreation ....this City Centre Plan new sets the pattern. Certainly, it implies that within certain principles, new development must conform. But this is negative thinking; the plan really shows the way in which positive and imaginative development can now take place (City Centre Planning Group, 1965, p1).

The plan was to provide a basis for: re-shaping the city's central area; the control of development; guidance on standards; public understanding and participation in planning the city centre (City Centre Planning Group, 1965, p3). Thus, the tone was set. This was to be a bold, ambitious, design-led plan for a major city; a plan that would be used not only to control but to provide a positive and inspirational model for others to follow. The ambition was to re-shape and redevelop what was perceived to be an obsolete and inefficient city centre. Whereas in the IPPS the importance of public participation was acknowledged, here it was raised to the status of one of the three main purposes of the plan.

Like the IPPS, the City Centre Plan comprised a first part analysing the existing situation, as shown in Table 3.1 below and a second part set out policy and proposals. But to this structure three additional parts were added: part three detailed how the plan was to be realised; part four reported on redevelopment currently in progress and part five included nine technical appendices. After setting the regional context, part one analysed city centre employment trends, the relationship with the port, topography and character, activities by sector (shopping, housing, hotels, etc), outdoor space, and each mode of travel. The whole was then brought together in an overview of 'the total problem'.

The main issue was thought to be that of the inherited structure of roads and buildings that in the past had been changed only fitfully and by piecemeal and uncoordinated actions.

> This lack of adaptation his in time brought obsolescence, unemployment and migration which have drained away or diverted investment skills and initiative from Merseyside (City Centre Planning Group, 1965, p53).

Thus was presented a clear picture of the planners' perceptions of the ills of Liverpool city centre: it was obsolete and congested. Many of the issues identified would not be out of place in a 21st century plan. Indeed it is revealing to find references to issues that some policy makers would claim to have 'discovered' in the 1990s, that were in fact recognised more than 35 years earlier. Examples included:

- encouraging a return to city centre living
- promoting cultural developments
- opening up the riverside
- increasing pedestrianisation
- reducing reliance on the motor vehicle.

Table 3.1   Problems and ambitions for Liverpool City Centre, 1965

The fabric of the city:	Obsolete buildings and transport arrangements must go. Fine individual buildings and groups of buildings must be preserved. The overall urban form should be clear, powerful and memorable.
Economy:	Need to create jobs for a growing population, to reduce unemployment and respond to the national shift towards the service sector. Need to replace 4 million sq ft outworn office space. Factories in unsatisfactory locations will need to be moved. Need to cope with the decline in warehousing.
Shopping:	Need to provide an attractive, re-organised and easily accessible centre.
Leisure and tourism:	Need to transform the existing range of hotels to meets the growing demands of business and tourism. Need to encourage cultural developments.
Housing:	Need to upgrade older council estates. 'More professional and middle-class households need to be attracted as residents in the rebuilt central area.'
Open space:	Need to provide more open space. 'The riverside should be opened up more so that it can be seen and enjoyed.'
The port:	Need to improve terminals for seagoing passengers and to re-use the sites of old docks that may become available.
Communications:	'The routes and timetables of trains and buses need to be co-ordinated.' New rail routes need to be opened up for commuters. 'Local and through (road) traffic need to be separated from one another and from pedestrian routes.' 'Congestion must be prevented by increasing the attractiveness of other means of transport.' There is a need for safe, attractive pedestrian routes to all parts of the central area.

Source: Liverpool City Council, 1965

Part two began with a response to the perceived need to provide 7.5m sq ft of office space over the next twenty years. The role to be played by office development was said to be critical because office employment was amongst the fastest growing sectors of the economy, an intensive user of land and often best located in the city centre (City Centre Planning Group, 1965, p59). Offices were to be developed in the Old Hall Street, Moorfields, Lime Street and Central Station areas in addition to the historic commercial centre around Castle Street. Significantly these zones, which would accommodate the highest densities of employment on Merseyside, adjoined and were integrated with the proposed loop and link underground railway stations. Industry and warehousing was to be banished to the edges of the city centre on sites adjoining the primary road system. 'Offensive' industry was to be removed from Old Hall Street, Dale Street and other inappropriate locations. A series of new shopping centres were under construction or planned. St John's Precinct, Central Station and Clayton Square have all been completed according to the plan. A fourth and in many ways the most interesting scheme, the Strand-Paradise shopping and entertainment complex, has never yet been realised in its original form although the area remains to the present day the subject of controversial and fiercely debated proposals.

The Strand-Paradise Street site contained a very large proportion of cleared land, was strategically placed between the business and shopping districts with excellent access from both the primary road system and the local rail system. There were magnificent views towards the Mersey from much of the site. The idea behind the scheme was to extend the city's shopping centre to meet modern requirements and to integrate this expansion with housing, leisure and other uses, all with convenient access from public transport. The scheme was to contribute to the townscape of the city by providing a landmark hotel tower on the Church Street/Paradise Street corner and buildings positioned and orientated to maximise views of the Mersey. Detailed proposals, originally derived from a 1963 consultants report (Shankland G, 1963), comprised three main elements:

- a large multi-use building complex behind and parallel to Lord Street at the northern perimeter of the site, accommodating a shopping, entertainment and exhibition centre, a hotel, 600-800 dwellings, a new central bus station linked by pedestrian subway to James Street station, car parking and ancillary uses
- a block, mostly on the former Customs House site, containing warehousing and service industry

- a park of 6.5 acres (2.6 hectares) between and defined by these two major elements.

Sadly, little of this bold proposal was developed as originally intended. On the main site a bus station and large multi-storey car park were built in the 1970s but no major shopping, housing or other uses were completed. A hotel was built further south along Paradise Street fronting onto a grassed area running down to the Strand and on the Customs House site a mixed office/service industry complex was constructed but demolished in the late nineties. By the millennium there were competing proposals for the redevelopment of the area.

The City Centre Plan also proposed that all principal shopping streets should become traffic-free in order to provide opportunities for new paving, street furniture, fountains, tree planting, kiosks, cafes and suchlike, to achieve a lively and varied environment (City Centre Planning Group, 1965, p65). Culture and entertainment were important elements in the policy but little was actually implemented. The important proposals to provide the city with an exhibition hall for industrial fairs and trade shows has never yet been realised; neither has the proposed arts centre, nor the sports centre, nor the youth centre.

Amongst the most interesting proposals was the idea of re-introducing housing into the central area. It was stressed that this would not displace other dwellings but would provide homes for those working in the centre. If the appearance, cleanliness and safety of the area were to be improved it was argued that residents of all classes might be attracted. On this basis the Strand-Paradise development was designed to provide 600–800 dwellings for middle and upper class residents and the river frontage between the Pier Head and Kings Dock 'could form the setting for some of the most desirable housing on Merseyside'.

> In addition developers would be allowed to provide high quality housing throughout the shopping and office areas of the city. To encourage them to do so they will be allowed to build dwellings over and above the permitted building density for the area, where appropriate (City Centre Planning Group, 1965, p69).

There were ambitious plans for open space provision. Four kinds of open space were thought necessary. Firstly, there was to be a widespread system of relatively small and intimate paved squares and courtyards. Secondly, there was to be a park that would provide 'an oasis of quiet and restfulness at the heart of the city'. Incredibly the area suggested for this park was that bounded by North John Street, Dale Street and Sir Thomas

Street at the heart of the modern office and restaurant quarter. Such a proposal clearly put great faith in the Council's implementation powers as it posed enormous problems with regard to the cost of land acquisition and the displacement of existing uses in the commercial core of the city. However, it was argued that it was the area most remote from primary road and the railway stations so not suited to high-density development. Promenades that exploited the 'views and atmospheres of the waterfront' were to be the third kind of open space, whilst the fourth was to be the provision of settings for formal buildings and 'a theatre for civic occasions'. William Brown Street and St Georges Plateau provided a fine example of this type of space. Parks were also to be provided around new housing areas and a series of connected open spaces were to link the core of the central area with the Anglican Cathedral, over a mile to the east.

Communications were a major concern of the plan. The city centre was to be encircled by an inner ring road, much of it built to motorway standards. It was argued that the purpose of this road was to take through traffic out of the centre and free existing streets to cater for local journeys, including the faster movement of buses, and to facilitate pedestrianisation. With a direct link from the national motorway network via the M62 this was a hugely expensive proposal that was not only going to be destructive of property but also create a major physical barrier between the central area and the rest of the city. At the time it was thought that the benefits in terms of traffic circulation and environmental improvements within the city centre outweighed the costs of acquisition, construction and physical intrusion.

It was proposed to increase the number of off-street parking spaces from the 1965 total of around 6,000 to 27,000 by 1981 of which about three-quarters would be directly accessible from the inner ring road without intruding onto the local street system. This number of parking spaces was said to be limited by the capacity of the primary road system and fell far short of estimated demand. The Council's policy was that excess demand would be channelled onto public transport and park and ride schemes. Thus despite the heavy investment proposed in road building, traffic restraint and the promotion of public transport were key elements of policy.

The desire to improve the pedestrian environment led the planners to propose an extensive system of segregated pedestrian routes through the city centre. In some parts of the centre (Bold Street, Church Street, Lord Street, Derby Square) this was to be a ground level scheme with horizontal separation from vehicular traffic. In other areas (e.g. in the rebuilt Moorfields area and around Old Hall Street) there was to be vertical separation with pedestrian walkways and access to buildings at high level.

All these various proposals for economic development, housing, open space and communications were brought together in a series of chapters dealing with urban form, built heritage and 'city improvements' through which the planners expressed their physical vision for the future of the city centre. The plan was therefore the product of the application of artistic principles as much as social science and economic analysis. It went beyond land use zoning to include the manipulation of urban spaces and forms in a comprehensive plan that would not only achieve an efficient centre but also one of physical beauty.

Implementation

Most of the new office building took place in the period from the late 1960s to the early 1970s and was concentrated around the Old Hall Street area. The anticipated office boom failed to materialise in Liverpool in the way it did in Manchester, Birmingham and other competing cities. After the national economic collapse of 1974 there was little commercial appetite for investment in Liverpool and some sites (e.g. at Moorfields) have remained undeveloped until the present time. Furthermore, as time went on and the planning system appeared to weaken, competing locations were able to attract offices that might otherwise have gone to the city centre. Most notable amongst these competing locations were the former dockland area, later under the control of the Merseyside Development Corporation.

Retail investment has included four enclosed shopping centres (St Johns Centre dating from the 1960s; and Cavern Walks, Central Station and Clayton Square all completed in the 1980s). In addition there has been selective rebuilding and refurbishment of premises in the core Church Street area and more recently the conversion to retailing of former warehouses and other properties in secondary locations, notably in the Duke Street-Bold Street area. All in all the effect has been a rather piecemeal and unplanned series of interventions that have lacked coherence either in terms of their effect on the spatial form of the shopping centre or in terms of the range and type of accommodation provided. In consequence it could be argued that the shopping centre has become characterised by a surfeit of secondary retail accommodation in fringe locations and a shortage of high quality provision in primary locations.

Figure 3.3   Merseyrail underground station at Moorfields
Passengers enter at first floor level: a legacy of the proposal for vertical segregation of pedestrians and vehicles in the 1965 City Centre Plan

Source: author

Scarcely any of the housing envisaged in the plan was ever built and the central area increasingly became a housing desert through the nest two decades. However, by the nineties Liverpool city centre, like many others, experienced a renaissance of demand for 'living in the city' with a rapid increase in the supply of both converted and purpose built accommodation (Couch, 1999).

One of the biggest failures of the plan was in the provision of open spaces. Through the redevelopment process planners did manage to provide a small number of useful squares and courtyards within and between developments but the park between Dale Street and Sir Thomas Street was never built and the Strand-Paradise Street area was grassed over but was so inadequately landscaped that it scarcely merited the title of 'park'. There was no parkland around housing and no open space link to the Anglican Cathedral. The inner ring road was only partially constructed in a scaled down version as changes in public attitudes and cutbacks in public spending reduced the government's interest in such investments. Whilst only some of the parking garages to be accessible from the inner ring road were built, there have been a number of parking developments in the heart of the central area (such as those at Paradise Street, Moorfields, etc.) accessible only through the local street network and against the principles of the original plan.

Although the 'loop and link' underground railway system was completed in the late seventies much of the complementary development of the suburban railway network has yet to materialise. Whilst most of the ground level pedestrianisation has been realised to great effect, the implementation of the vertically separated high-level system proved a disaster. Only a number of short sections were ever completed: an area to the west of Old Hall Street was the most extensive and there were other short sections around James Street, Moorfields and Williamson Square. High level walkways proved unpopular for a number of predictable reasons: access was difficult; they offered little if any advantage over ground level access; and they became targets for graffiti and vandalism and were perceived as dangerous, especially at night. By the late 1970s the building of the high-level walkway system had ceased and by the millennium most of the outlying sections had been removed (notably around Williamson Square).

The vision of urban form espoused by the plan was also eroded as time went by. The proposals had been developed at a time when the concepts of 'redevelopment' and 'renewal' were dominant in planning theory. Much of the proposed development was illustrated by images of a redeveloped city centre based on modern architecture. During the subsequent decades public tastes in Britain turned against this style, whilst a changing economic climate and shifting political preferences moved planning policy away from large scale redevelopment towards cheaper piecemeal renovation and improvement schemes. In consequence, the proposed changes in urban form became discredited by association with modernism and the redevelopment process. At the same time the planners were becoming more concerned with the social and economic problems of the city and interest in urban design waned. Although conservation and the preservation of listed buildings was to become an important part of planning policy, by the eighties there was no longer any comprehensive overall vision of the future physical form of the city centre. With the onset of recession, the desperation to attract development of any kind considerably weakened the Council's ability to maintain high standards of design control in individual developments. Furthermore, the political agenda was changing with emerging concerns about urban deprivation and the mechanisms and policies required for its treatment.

Social Malaise

It was in 1966 that Liverpool City Council undertook an innovative social survey of inner city residents. Earlier that year the Council has approved the a massive slum clearance programme (see below) which meant upheaval for over 100,000 people, one in seven of the then population. In order to minimise this upheaval and distress the Council acknowledged that it was necessary to know more about the people who lived in the clearance areas. To this end a social survey was undertaken on behalf of the City Council. Over 3,000 households were interviewed and in an innovative use of information technology the analysis was, in part, undertaken on the City Treasury's new ICL 1904 computer (City Planning Officer, 1966, p1). The conclusions represented one of the earliest local authority discussions of inner city social conditions.

> The material gained from the survey pointed to the fact that the people to be re-housed were often those least able to cope with the problems that such a major upheaval as the inner areas redevelopment programme would necessitate. In all aspects of life – housing, employment and social class – it was predominantly a static population. This alone would be a barrier to coping with enforced change, but this barrier was reinforced by yet other characteristics exhibited by the survey population. Incomes were low, making affordable rents lower than actual rents of new housing. Standards of education and training did not equip the working population to tackle the problems of obtaining new jobs, many of which are in new light industries of the type attracted to new towns. People did want better housing conditions, but only half of the households were prepared to move to find them, and few were prepared to go beyond the City boundaries....not only were they generally reluctant to move, but they would be unable to pay for their better environment (City Planning Officer, 1966, p15).

Subsequently the Council commissioned further detailed area-based studies of social conditions 'to discover by study a better definition of inner areas and their problems' (PRAG, 1975, p17). Parallel with this burgeoning of social scientific investigation came a strengthening of corporate planning within the local authority and a period of action-research and experimentation in urban policy. In 1970 a review by

McKinseys, a firm of management consultants, proposed that the City Council committees and departments should be reorganised along corporate management lines with a chief executive given overall responsibility for the achievement of corporate aims. At this time the City Council itself was the agency responsible for the implementation of much urban policy. Three documents were published that were to indicate how the policies of the IPPS and City Centre Plan would be translated into corporate (i.e. City Council) programmes and implementation. These documents were the City Centre Plan Review; the Inner Areas Plan and the Outer Areas Plan. It was to become part of the corporate planning process that these documents would be reviewed annually to update programmes and respond to changing circumstances. Each plan was reviewed annually until 1974 when, as a consequence of the Local Government Act 1972 the County Borough of Liverpool was replaced by the Metropolitan Borough of Liverpool, albeit within the same geographical boundaries but with a somewhat different legal status, political leadership and staffing. Two of the most important of these plans are considered here: the Inner Areas Plan 1970 and the Liverpool Inner Areas Plan Review 1974.

Inner Areas Plan, 1970

The Inner Areas Plan (IAP) was concerned with the mixed zone of residential and industrial districts to the east of the docklands and city centre. These included Vauxhall, Everton, Kirkdale, Kensington, Edge Hill, Abercromby, Toxteth and the Dingle. The IAP was not a full restatement of planning policies, for it said little about development control, but a statement of corporate planning policies and programmes. It contained very little by way of analysis or policy justification, as that had already been provided in the earlier IPPS; nor much consideration of financial constraints or problems of implementation, as they were intended to be dealt with in the more detailed plans that were to follow. Thus the IAP and its sister plans for the city centre and outer areas, formed part of a chronological and spatial hierarchy, beginning with the Liverpool Interim Planning Policies Statement of 1965, which was elaborated in these plans for the Inner, Outer and City Centre areas, and was then to be translated into action on the ground through a series of 'action area plans'.

As with the separation of strategy from tactics that was evident in the IPPS, so the notion of 'action area plans' also had its origins in the PAG report and the 1968 Town and Country Planning Act. The thinking was that these detailed plans would set out a detailed physical planning framework and programme for the intensive and comprehensive redevelopment of individual areas. Supporting this statement of policies were appendices that provided detailed information on programming and scheduling of individual housing redevelopment sites, road construction schemes, primary and secondary school building, further education provision, open space and playing field provision, industrial sites and a schedule of district centre proposals. Thus the role of the IAP was not only to translate strategic policy into a spatial planning reality but also to facilitate the co-ordination of the development programme. Given that it was anticipated that much of the plan's implementation was to be undertaken by the local authority itself, the plan was an important corporate planning document and significantly, the planning department had acquired a key role in the Council's new corporate management structure.

Table 3.2. indicates the chapter headings in the IAP and illustrates the scope of the plan and the importance of the local authority itself in the implementation process.

Table 3.2   Implementing the Inner Areas Plan

Chapter	Policy	Main agency for implementation
Housing and population	Clearance and council housebuilding	LCC
	Housing improvement	LCC / private owners
Circulation and transport	Roadbuilding etc.	LCC
	Pedestrian circulation	LCC
	Public transport	MPTE
Education	School provision	LCC
	Further Education	LCC
Open Space	Open space provision	LCC
Industrial Sites	Industrial sites provision	LCC
District and local centres	Shopping facilities	Private sector / LCC
	Social facilities	LCC

LCC   Liverpool City Council
MPTE   Merseyside Passenger Transport Executive, created under the provisions of the Transport Act 1969

Source: author's analysis

Back in 1965 the National Building Agency (NBA) had reported on housing needs. Following their recommendations the City Council had agreed in 1966 to a massive clearance programme. In all 78,000 dwellings were to be demolished: 36 per cent of the total city housing stock but more than 70 per cent of the dwellings in the inner areas. By the time of publication of the IAP about 17,000 dwellings in the inner areas had been cleared under this programme, with a further 32,500 scheduled for future demolition. This was a massive programme. Those intended for clearance in the later stages of the programme, with a life expectancy of at least 15 years, were to benefit from renovation grants. The Housing Act 1969 had permitted local authorities to designate General Improvement Areas (GIA) within which housing renovation and local environmental improvements would be combined as an alternative to slum clearance and rebuilding. Initially two General Improvement Areas were designated at Kensington Fields and Granby, although many more designations followed over the next decade.

> A combination of circumstances led to a reduction in planned residential densities compared with the original plan. In 1965 the average net density envisaged was for 140 persons per acre rising to peaks of 200 ppa around district centres and transport nodes. But whereas it had then been anticipated thay the population of the inner areas would fall from 266,000 to 156,000 by 1985 (an average reduction of 5,000 per annum) the actual rate of decline was probably twice this figure. In consequence there was less need for such high densities. Furthermore, by 1970 both national and local politicians had turned against high-rise housing and a more humane scale of housing was being proposed. Thus the IAP envisaged that the attainment of good housing environment with a satisfactory balance of dwelling types is possible at a density range of 90 to 120 bedspaces per acre. Within this range development need not exceed 4 storeys in height (Liverpool City Planning Department, 1970, p2).

The IAP paid little attention to public transport, indicating that this was now primarily the responsibility of the MPTE. Road construction, on the other hand, was a major concern. The schedule lists a large number of major road-building projects including the inner ring road, radial connections through the inner areas; the inward extension of the M62 motorway; and tangential routes including the Low Hill distributor. There is no equivalent schedule of pedestrianisation or traffic calming schemes, although it is said that these would be dealt with in the more detailed action area plans.

The provision of educational facilities was another major concern of the IAP. Once population forecasts had been made, formulae were applied to calculate the need for schools of various types, sizes and locations. There was a great deal of apparent certainty in this exercise as on the one hand, once children were born their educational needs were reasonably easy to predict so long as migration patterns remained stable, whilst on the other hand the City Council and the churches were the only major providers of schools. Most schools had fixed catchment areas and there was little parental choice with regard to which schools their children were to attend. This was an era when the dual use of school buildings and playing fields, for both educational and community use was much in favour. Sharing playing fields allowed for economies in provision, especially in the inner cities where open land was in scarce supply. Using school buildings for community activities was seen as one mechanism for strengthening links between school and community and a way of bringing inner city parents and others into the ambience of educational facilities.

In line with the planning ideology of the time, the IAP was concerned to ensure that the quantity of open space provided across the city was in accordance with national standards; these being in the order of 6 acres[1] per 1,000 population. This provision was to include 1 acre for ornamental open space and 2 acres of playing fields to be within ¼ mile of the place of residence, together with a further 3 acres elsewhere in the city. Recognising the two issues of the environmental intrusion of roads and the emerging problem of derelict land, much new open space in the inner areas was to be provided as a 'buffer zone' between primary roads and buildings so as to reduce noise levels or as an 'interim use' for cleared sites so as to reduce unsightliness.

Industrial sites were to be provided both for the relocation of firms affected by redevelopment schemes and for new industrial development. Within the inner areas surplus railway land provided the biggest source but total supply fell short of predicted demand by about 50 per cent. The remainder was to be provided on the outskirts of the city. Thus although it was estimated that some 50–60 acres of industrial land was needed to meet the needs of the inner area population, about half was to be provided elsewhere. Such a shift in the location of employment would inevitably encourage further outward migration and exacerbate the depopulation of the inner urban areas.

Provision for district centres was to follow the proposals of the IPPS. Four such centres were proposed within the inner areas at Park Road, County Road, Breck Road and Wavertree Road. The last named was a typical example. It was anticipated that there would be a requirement for 130,000–150,000 sq. ft. gross shopping floorspace. At the time Wavertree Road between Overton Street and Tunnel Road was already a focus for

---

[1] One acre is approximately 0.4 hectares.

shopping in the Edge Hill district: it contained Freemans department store, Woolworths, and a number of other multiple retailers. It was proposed to redevelop most of the shops on the south side of Wavertree Road to provide new retail units together with community facilities including a police station and social security office. The small, older shops on the north side of the road would also be retained. Traffic was to be diverted away from Wavertree Road onto a new diversionary route. What was not anticipated by the policy was the subsequent rapid decline in local population and the consequent reduction in local retail spending. Neither did it anticipate the changes in shopping provision whereby retailers were seeking ever larger but fewer outlets for their goods, nor the greater mobility and retail choice brought about by rising car ownership, even amongst inner area residents. As a result the existing department store and other multiples closed shortly after the plan had been published, Wavertree Road was not closed to traffic; the proposed retail investment never took place in the form intended, although a small retail park and discount supermarket were developed on nearby land at Edge Hill. By the year 2000 the title of district centre was no longer being applied to the area.

Liverpool Inner Areas Plan Review, 1974

Four years on and the corporate investment plans for the inner areas were being informed by revised population projections, growing concern about the speed of policy implementation, and emerging worries about the local economy and social deprivation. Despite experiencing population decline of 23 per cent between 1966 and 1971 the City Council were still confidently predicting that:

> After 1976, the population is likely to increase as the level of demolition reduces and more families are re-housed within the Inner Areas as a result of the current building programme concentrating resources on redeveloping inner area sites (Liverpool City Council, 1974, p7).

Strains in the housing redevelopment programme were beginning to appear. Re-housing was taking longer than anticipated, not least because families were becoming increasingly selective about alternative accommodation. There were a number of reasons for this: firstly, the post-war housing shortage was on the wane and the relationship between housing demand and supply slowly being tilted in favour of consumers. Secondly, a growing number of households were living in the new replacement housing (both inner city multi-storey accommodation and distant overspill estates) and the reported experiences of those who had moved were not always very favourable. Finally, the other option of re-lets in existing council housing was increasingly rejected, especially in the inner areas where much of the property was perceived as obsolete and offering little advantage over insanitary but familiar slums. This last problem also affected the ability of the Council to re-house people near to their former homes in the inner areas and a higher proportion than originally anticipated was being re-housed through overspill. This in turn fuelled community dissatisfaction and further depopulation of the inner areas.

However, the Council was moving away from large-scale slum clearance programmes and by 1972 had approved ambitious plans for improving 61,000 older private sector dwellings in the inner areas (nearly 30 per cent of the city's housing stock). The obsolescence of the older public sector stock was also being recognised and a refurbishment programme begun. The question of housing dissatisfaction amongst council tenants was becoming a matter of such concern that the Director of Housing undertook a survey in conjunction with the University of Birmingham in the hope of revealing the causes of 'difficult to let' properties, housing preferences and tenants' views on the design of new dwellings.

The attitude towards housing density and form had also changed since the IPPS in 1965. Popular reaction against multi-storey living was having a political impact and the Council's policy was to be 'more flexible'. Henceforth, all development was to be subject to a maximum height of three storeys. Indeed, from January 1973 all new municipal housing schemes were to be limited to only two storeys (Liverpool City Council, 1974, p9). However, some of these schemes still managed to attain densities of 210-235 bedspaces per hectare. Many of these high-density low-rise schemes were to prove even less popular than the high-rise alternative. The most notorious, such as the Radcliffe and Brunswick estates lasted little more than a decade: a scandalous waste of public resources.

Economic performance and unemployment were also emerging concerns for the City Council. The unemployment rate in the inner areas was about 30 per cent above the city average which was itself almost double the national average. Although acknowledging that the most significant influence on employment was the national economic situation, it was noted that the recent rationalisation of port activities was having a substantial impact on employment in the docks and in port-related activites such as transport and warehousing. The relocation of firms in search of

modern premises or more extensive sites on the periphery of the
conurbation was also impacting on workers in the inner areas who faced a
choice between longer and more expensive journeys to work or
unemployment (Liverpool City Council, 1974, p11).

A review of transportation policies had been undertaken in 1973.
Policy was still strongly in favour of developing public transport provision
within the city.

> Current policy recognises that the rate of road construction is restricted by
> financial resources, and concludes that if environmental areas free of through
> traffic are still to be created the current shift from the use of public transport
> to the private car for peak hour journeys-to-work to and from the City Centre
> must be halted (Liverpool City Council, 1974, p13).

Some proposals had reached an advanced stage of development. For
example there were proposals for two new bus mobility schemes: the first
was for a bus way as a new carriageway on the west side of Park Road,
superseding the previous proposal for a dual carriageway for all traffic; the
second was on Great Homer Street where one lane of each carriageway of
the proposed dual-carriageway would be for the exclusive use of buses
(Liverpool City Council, 1974, p13). However, in April 1974 most aspects
of transportation policy became the responsibility of the new Merseyside
County Council and in the light of other priorities, neither scheme was
implemented.

With regard to road proposals, whilst acknowledging the slow rate of
highway building since the IPPS, and despite growing public and political
debate about the urban road schemes, the Inner Areas Plan Review
continued to list most of the major road proposals as being in various stages
of preparation. It was merely stated that the responsibility for
transportation planning would pass to the new Merseyside County Council
in April 1974 who would take decisions about the implementation of these
schemes.

In education policy, the declining number of school age children had
become a major factor in the inner areas. At the same time many school
buildings were obsolescent and suffered deficiencies in the provision of
playing fields and other facilities. This put the local authority in a difficult
position, for there was little point in investing in schools that might have to
close due to falling rolls, and rationalising school provision was a highly
sensitive and political issue. Furthermore:

> the complex pattern of social and environmental problems, and their close
> links with educational deprivation, have led to the classification of a

substantial proportion of the Inner Areas as educational 'areas of need', qualifying for extra resources in terms of staff, books and materials, and maintenance (Liverpool City Council, 1974, p17).

These were part of the 'Educational Priority Areas' policy under which the Department of Education and Science had designated a number of selected inner city districts, within which additional funds would be aimed at improving educational provision and attainment. It was argued that the educational performance of these children was below average because of a 'cycle of deprivation' in which their disadvantaged parents and peers placed a low value on education and offered little extra-curricula support. By breaking this cycle through the education system it was argued that such children might achieve better jobs with higher incomes and escape from the inevitability of continuing poverty. A detailed account of the Liverpool Education Priority Area project can be found in Midwinter, 1972.

The 1974 review also commented upon the provision of social and medical services. It was only since 1968 that the city had had a unified Social Services Department and its field services were now being decentralised to district offices. One concern of the plan was the proper location of these offices as well as other forms of community and residential provision. Although the reorganisation of the National Health Service was due to take place in April 1974 the review gives some consideration to the provision and location of hospitals and other health services within the inner areas.

Environmental quality was said to make an important contribution to the 'character and quality of the inner areas as a place to live and work'. A list of local environmental improvement policies being employed by the City Council at the time is shown below:

- comprehensive environmental care and maintenance, the Brunswick Neighbourhood Scheme, for example, was benefiting from an intensive programme aimed at regular street sweeping, removing rubbish, treating vacant sites, tree planting, repairing paving slabs and roads
- environmental improvements within industrial estates, such as Brasenose Road where a scheme included landscaping, cleaning buildings and improving fences, hoardings etc.
- a special environmental assistance scheme (Operation Eyesore) had provided some central government subsidy to provide temporary work for the unemployed on environmental improvement schemes

- smaller clearance areas – by the seventies the size of clearance areas had been sharply reduced and it was hoped that this would reduce the amount of unsightly cleared sites awaiting redevelopment
- the rehabilitation of the inter-war council estates within the inner areas
- improvements in the cleansing of all public areas
- the appointment of district environmental officers to co-ordinate action at local level
- using development control to improve the quality of building design and finish
- conservation – controls over areas of special architectural and historic interest and important groups of trees was gradually being extended (Liverpool Inner Area Plan Review, 1974).

In response to growing pressures from the environmental movement, the Civic Amenities Act 1967 had, inter alia, provided local authorities with powers to extend the protection of the built environment from individual buildings and structures to areas and groups of buildings of special architectural or historic interest. The mechanism for this protection was the 'conservation area'. Liverpool City Council was soon amongst the most active authorities in this field and by the time of the 1974 Review had declared eight conservation areas. Two were in the city centre: Castle Street (including the Pier Head) and William Brown Street. A further six formed a continuous link between Rodney Street and the Canning Street area, through Princes Road to Princes Park, Sefton Park and Mossley Hill. There were two further conservation areas at St Michaels Hamlet and Fulwood Park. Apart from those in the city centre virtually all of the conserved areas comprised middle and upper class residential environments from the 19th century. At the time of their construction, the specification and aesthetic quality of these areas contrasted sharply with the poverty and squalor of the working class districts of Victorian Liverpool.

The hierarchical approach to retail provision and the commitment to building up district centres would continue. However, it was now recognised that changes in the retail economy were beginning to threaten the viability of this policy, notably through the trend towards larger freestanding supermarkets and hypermarkets.

> Hypermarkets clearly do not meet the objectives of Liverpool's shopping policy. In particular, although they may offer cheaper prices they erode the viability of the 'high street' or district centre shops, and of public transport,

upon which a substantial minority will always depend... The Council policy is not hostile to large units as such...provided they are located...at district centres or in the City Centre (Liverpool City Council, 1974, p26-27).

The review also recognised the idea of 'priority areas', acknowledging that many central and local government initiatives had recently been characterised by an attempt to direct resources to priority areas of need. The theory supporting such a policy was that 'positive discrimination' in favour of areas that exhibited social, economic and environmental problems would go some way towards solving those problems. By the 1970s Liverpool was the recipient of a number of area based initiatives and research studies sponsored by central government and other agencies. These included the Home Office's Urban Programme and Community Development Projects; the Department of the Environment's Inner Areas Studies as well as independent initiatives such Shelter's Neighbourhood Action Project.

Conclusions

One of the key features of these plans was just how important they appeared to be to the policy making and political ambitions of the City Council. There is a strong feeling that these plans played a central role in the corporate planning of City Council spending. This can be seen particularly well in relation to the IAP and its role in co-ordinating policy and investment in the inner areas. The ability of development plans to drive corporate planning was facilitated by the fact that the Council itself still retained responsibility for the great majority of local public services: the era a fragmentation and privatisation had yet to arrive. Table 3.3 below illustrates these and other characteristics of planning and urban regeneration during this period.

Another feature that differentiated these plans from their predecessors was the idea of viewing the city as a system of interconnected elements and the recognition that changes in land use would have impacts upon transportation. The IPPS took a holistic view of the city and in a key innovation: strategy was separated from the tactics of local planning and implementation.

Table 3.3   The characteristics of planning and urban regeneration in Liverpool during the period of modernising the city

What was the relationship between planning and urban regeneration?	A strong commitment to planning within the City Council. Urban regeneration programmes are contained within plans and planning policy.
What were the main achievements of planning and regeneration during the period?	A comprehensive and decisive set of plans were produced. Large-scale slum clearance and rebuilding was achieved. There was substantial investment in physical infrastructure and the city centre.
What was the relationship between economic, environmental and social aims?	Strong emphasis on physical environmental improvements and the creation of a modern and 'efficient' urban structure. Economic development was an important aim of regional policy but of limited local planning concern.
What was the extent of local democracy and participation?	Little direct action by central government. The City Council was relatively free of central government intervention in local policy. Little community participation in local planning and policy making.
To what extent were policies co-ordinated as part of long-term strategy of fragmented and short-term?	Rational comprehensive planning. Clear relationships between strategic (conurbation and city-wide) planning and local area plans.

Source: author

The main aim of these plans was clearly to modernise the physical form and structure of the city, and in a fairly radical way. With little regard for cost, new highway and public transport systems were to be built and obsolete districts torn down and rebuilt. In this regard the plans were over-ambitious. Firstly, because it must always have been clear that the cost of proposals and the source of funding had not been sufficiently thought through. Secondly, because the plans were based upon assumptions of continued economic and population growth that very soon turned out to be grossly over-optimistic about the future fortunes of the local economy. The consequences of this over-ambition could most readily be seen in the swathes of land acquired for urban motorway systems and in some of the physical changes that were begun in the city centre, for example building of the high-level pedestrian walkways.

Transport was a major concern of these plans. The IPPS and City Centre Plan were dominated by transport strategy and investment proposals. Reflecting their modernist view of the world, these plans did not put a great deal of emphasis on the protection of the city's built and natural heritage – although this was later to become a key element of city planning policy. Yet in some fields these plans showed themselves to be ahead of their time: the IPPS recognised the importance of public 'ownership' of the plan and of public participation in the plan making process; housing in the city centre was to be developed; investment in public transport was to be encouraged and traffic flows were to be controlled through parking policy. Many of these were innovative and radical ideas.

By the beginning of the seventies the Council had begun to recognise the emerging social, economic and environmental problems of the inner areas. The slum clearance juggernaut was gradually put into reverse. Multi-storey blocks were no longer to be built by the Council. Area improvement became the dominant policy for treating obsolete housing and new experimental policies were targeting the inner areas with additional resources to tackle urban deprivation. These policies are the concern of the next chapter: deprivation and the inner city.

4 Deprivation and the Inner City

By the late 1960s the scope of British town planning was being broadened to embrace social and economic concerns as well the area of physical development that had been its traditional strength. This change was most noticeable in the bigger cities, including Liverpool, where there was a growing concern that urban policy should be seen to be more responsive to the economic and social needs and desires of local people. Within a decade Liverpool had become the recipient of a number of area based initiatives and research studies sponsored by central government and other agencies. These included the Home Office's Urban Programme and Community Development Projects; the Department of the Environment's Inner Areas Studies as well as independent initiatives such Shelter's Neighbourhood Action Project. Figure 4.1 indicates the distribution of these experiments, studies and programmes within Liverpool.

The Urban Programme

The country had moved from the relative complacency of Macmillan's 'never had it so good' to the discovery that 7 million people (14 per cent of the population) lived in poverty (Abel-Smith and Townsend, 1965), and that there still remained over 1.8 million unfit dwellings countrywide (MHLG, 1967). Furthermore, there were concerns about slum clearance policy and calls for more housing renovation (Gibson and Langstaff, 1982, Ch3). Urban dereliction was a growing problem as industrial restructuring and technological changes led to the abandonment of many inner city industrial premises, railway yards, docklands and so forth. By 1971 it had also become clear that the major cities were losing population at an alarming rate. Social unrest and community breakdown were becoming apparent in some areas. There was clearly an emerging 'urban' problem different in character from any that had previously been experienced in Britain.

Nationally, the first major development in 'urban' policy came in 1968 when the Government was forced to respond to social unrest caused by poor race relations and growing urban inequalities by establishing the 'Urban Programme'. In the words of the Home Secretary:

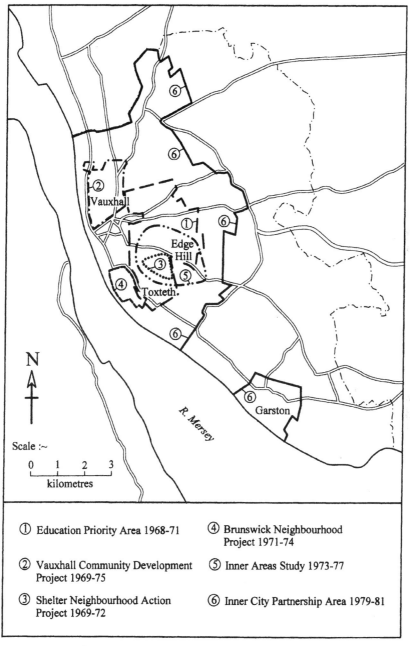

Figure 4.1   Inner city experiments, studies and programmes, 1969-79

Source: author

> There remain areas of severe social deprivation in a number of our cities and towns – often scattered in relatively small pockets. They require special help to meet their social needs and to bring their services to an adequate level...The purpose is to supplement the Government's other social and legislative measures to ensure as far as we can that all our citizens have an equal opportunity in life (Gibson and Langstaff, 1982, p147).

However, as the policy response came from the Home Office, the department responsible for law and order, it might be construed that the main concern was to tackle social unrest, rather than any deeper concern for social inclusion. Nevertheless, the programme provided selective, if small scale, subsidies to local projects such as nursery schools, advice centres, community centres and language classes for immigrants (Gibson and Langstaff, 1982, p147).

The programme, which became known as *Urban Aid,* allowed local authorities to bid to the Home Office for a 75 per cent grant to subsidise the cost of approved projects in deprived areas, with the remaining 25 per cent being found from local sources. Projects were small in scale with an average cost below £10,000 each at current prices.

In Liverpool the programme was taken up with some vigour and by 1974 more than 50 projects had been completed or were under way across the inner areas. These included the provision in the inner areas of a number of nursery classes, day nurseries, adventure playgrounds, community centres, housing and legal advice centres, care facilities for the elderly; a children's home, all-weather sports pitches, and a variety of other small scale projects (Liverpool City Council, 1974, appendix 12).

Although the projects were small there was some evidence of a strategy guiding their location. Funds were spent in areas of 'special social need'. Using information from the Social Malaise Study and other sources, the City Council had identified areas of the city that were experiencing the worst multiple deprivation and where action was to be prioritised. The emphasis of urban programme expenditure in Liverpool was on education, especially at the pre-school stage, youth activities and welfare advice. Unemployment, even in Liverpool, was not seen as a major problem until the early 1970s and certainly not as a significant cause of deprivation. The causes were generally thought to be social, resting within the community and the individual. The provision of education and advice would enable 'the deprived' to better themselves while playgrounds and organised activities would keep youth 'out of mischief'. Such an approach to policy reflected a social pathology and cycle of deprivation approach to urban social problems.

There was some evidence of co-ordination with other policies. Resources provided by the urban programme were used to complement other funds in the provision of facilities within the new 'general improvement areas' and special funds were allocated to a 'priority neighbourhood scheme' in the Brunswick area. The Brunswick Priority Neighbourhood Scheme was a three year experiment that started in 1971, again 75 per cent funded by the Home Office. Its aims were to assess community needs, promote community development and the integration of local services. In support of these aims the City Council undertook community consultation and participation exercises and developed a number of schemes including the provision of facilities for youth and the elderly, traffic management, the landscaping of vacant sites and other remedial environmental action (Liverpool City Council, 1974, p28).

Vauxhall Community Development Project

Sandwiched between Everton and the industrial areas behind the north docks lies Vauxhall, one of the most deprived inner city residential wards in the country. Traditionally the area was home to people who worked in the docks and port-related industries such as food processing but with the technological and industrial changes of the 1960s and 1970s these jobs began to disappear at an alarming rate. Between 1967 and 1972 more than 20,000 jobs were lost from the area. By the end of the decade unemployment stood at over 36 per cent, more than twice the conurbation average (McIntyre B, 1995, pp43-48).

The Social Malaise Study had already shown that Vauxhall had severe problems. The area was therefore an obvious candidate to be one of the new 'Community Development Projects' (CPD) sponsored by the Home Office. These action-research projects were established in 1969 in order to obtain a greater understanding of the nature and causes of the multiple deprivation that affected so many older industrial areas. In seeking solutions the projects were to put considerable faith in the then novel idea of community participation: encouraging local people to participate in the policy making process. Even nationally the funding was modest: some £5 million shared between twelve local authority areas. Anticipating that the solutions to deprivation were likely to be found locally, each project was supported by a local action team that would promote community development. These teams were to undertake research into the nature of the problems facing the area and undertake some policy evaluation

(Gibson M and Langstaff M, 1982, p148). Initiatives developed by the Vauxhall CDP action team included:

- a study of local employment conditions
- appointment of a Department of Health and Social Security liaison officer at a multi-service centre (one-stop shop)
- community education programme, education resource centre and reading scheme
- resident-run information centre
- neighbourhood newspaper
- support for community groups

This localised approach to urban deprivation, aimed at community development and improving the delivery of services, was welcomed by the community even though it quickly became apparent that such a small-scale initiative would achieve little.

> To many people in Vauxhall it was evident that these schemes were doomed from the start...as they concluded that the problems had structural roots which lay beyond the designated boundaries of the selected areas. The situation in Vauxhall, it was claimed, was due to purely external restructuring and change...to which there was no local solution (McIntyre B, 1995, p56).

However, the research team, based at Oxford University, took a considered view. With regard to individual initiatives that comprised the Vauxhall CDP they commented that:

> Many of these initiatives, whatever their scrambled beginnings, have lasted to become part of the local scene – the Scottie Press, the Information and Law Centre, the under-13 football league, the community centres, the large number of resident and issue based groups, and the debating forum of the neighbourhood council, which succeeded the project steering group...The area benefited from a programme of environmental improvements; derelict sites were grassed, and trees planted; several play areas were created, a swimming baths and wash-house due for closure were retained. Plans for housing modernisation were brought forward...Some of the major roads running through the area were more strictly segregated from pedestrian areas, and a traffic management system introduced within the area (Topping P and Smith G, 1977, p114).

There were some very positive long-term outcomes from the CDP, such as the community-based Vauxhall Neighbourhood Council (VNC) and the neighbourhood newspaper. On the broader question of the value of the overall programme, the research unit also came to the view that the changes affecting Vauxhall were of a structural nature and that the recent industrial closures and downturn in employment were no more than the latest in a long series of events that maintained Vauxhall's seemingly permanent position as one of the most deprived wards in the city. In a telling analysis of the economic problems facing inner urban areas the authors suggested that:

> One of the major problems we face is the orderly regeneration of the older industrial areas such as Vauxhall. Here the comparison with primitive agricultural methods – where areas were cleared, planted, burnt out and abandoned for new sites, is instructive. In time new methods allowed continuous use of the same areas. Yet in contrast the process of industrial change is quickening: areas are burnt out more rapidly, and social capital is wasted. Clearly we need to move away from this destructive pattern; but the problem is how to achieve this, particularly in areas like Vauxhall where the run-down has already gone so far (Topping P and Smith G, 1977, p120).

Whilst the Vauxhall CDP action team may have been aware of this view, they chose to concentrate on ameliorating this situation through local community development and arguing for improvements in the delivery of local authority services. (Topping P and Smith G, 1977, p119). Nationally, some CDP teams were taking a more radical view, especially as it became apparent that there were similar problems being faced by each of the twelve areas. Gradually the view was formed that these problems were essentially symptoms of structural changes in the wider economy, the workings of the housing system, and of the health, social and education services. By 1974 the National CDP Inter-Project Report was concluding that:

> The problems in these areas are not going to be solved by marginal rearrangements to take account of their special minority needs. From its small base, CDP can map the points at which private and public policies are having negative and unequal effects. We can also aim to explore strategies for developing local awareness of these points and for raising them to greater public attention. But the major response must come from central and local government, with substantial changes in policy and the allocation of new resources (National CDP, Inter-Project Report, quoted in Gibson M and Langstaff M, 1982, p154).

In the south end of the city Granby ward presented a very different picture of inner city deprivation from Vauxhall. This was a chronically deprived multi-ethnic and multi-cultural neighbourhood that had traditionally provided the first home to many inward migrants to the city. The population was more transient than in Vauxhall and contained a much higher proportion of students (almost unknown in Vauxhall at that time) and younger childless households. The housing was different too, with more larger, older dwellings in multiple occupation, more private housing and less council owned property. It was this district that Shelter selected for its 'neighbourhood action project'.

Shelter Neighbourhood Action Project (SNAP)

Shelter, the national campaign for the homeless, had been established in 1966 and by the late sixties felt the need to move beyond campaigning to develop an action-research project in order to learn more about the causes of and solutions to housing deprivation. Based in Granby, the project ran from 1969 to 1972. Area regeneration and environmental action provided the initial focus for action. By 1972 a combination of local council activity and SNAP support within the community had made Granby a successful General Improvement Area (GIA). SNAP did not seek to attach blame to individuals for the situation that they found in Granby, rather they argued:

> In our experience in Liverpool, apart from the 'Rachmans' and the more vicious exploiters of the poor, there are no villains. The public administration of poverty appears to be benign and, when it comes to any individual manager or public servant, undoubtedly is benign. The basic complacency derives from distance. In few major social problems is there such a distance between those affected and all those who, conceivably, could do anything about it. It is possible, alas, to carve out a career studying the problems of redundancy! The people of the inner areas have been surveyed, studied and written about until the whole process has become part of their tragic experience (SNAP, 1972, p212).

Nevertheless it was clear to SNAP, as it had been to the researchers in Vauxhall, that the area suffered multiple deprivation caused by external forces beyond the control of the local community and that this would not be solved by local housing and environmental improvements. SNAP rejected the social pathology explanation of urban deprivation but argued instead that the area had to be seen as part of the wider urban economy and that the fundamental causes of deprivation lay in structural changes in employment and housing markets. It was also clear that these changes were reinforced by processes of suburbanisation aided and abetted by slum clearance and dispersal policies (Gibson and Langstaff, 1982, p149).

SNAP concluded that it was beyond the financial ability of cities such as Liverpool to solve the problems of urban deprivation. Accordingly they proposed that the bigger cities should receive a higher proportion of the Government's Rate Support Grant and that a new, larger, Urban Programme should be established for the purpose of regenerating the economic base of these cities. Additionally SNAP found institutional and managerial failings that were exacerbating inner city deprivation, in particular a lack of co-ordination between different parts of the local authority and between local and central government. In response they called for a more corporate approach and more local accountability in the planning and delivery of services, or in today's parlance – more 'joined-up thinking'. Thus SNAP, working in Granby, added another dimension to the analysis of the problems of inner urban areas. Between this approach and that of the CDP in Vauxhall, both geographically and intellectually, came a third study: the Department of the Environment's Inner Area Study, based in the Edge Hill district.

Inner Areas Study

Bigger by far than either of the previous projects was the Inner Areas Study (IAS). The Liverpool study was one of three commissioned by the Department of the Environment in 1973 (the other two being in Birmingham and Lambeth). The brief was to:

- establish a better definition of inner areas and their problems;
- test the utility of various 'actions affecting the physical environment' that would lead to social and environmental benefits;
- explore the concept of 'area management';
- provide a basis for more general conclusions and policy recommendations (Wilson and Womersley, 1977, p6).

Central to the study was the idea that there needed to be a 'total approach' to the inner areas.

The need for a total approach at the local level was confirmed by our initial discussions which covered a wide spectrum of issues. They ranged from the management of council housing to the problem of single parent families; from uncertainty about the future of individual streets to crime and prostitution; from the opportunities for adult education to street sweeping and litter (Wilson and Womersley, 1977, p13).

This approach was similar to that being adopted in plan making: a view of the city as a series of systems and an acknowledgement of the interaction between the many variables that make up the character of a locality. Improving the management of council housing might have some benefits but would not, of itself, impact on the poverty in which many households found themselves. Removing uncertainty about whether a particular street was to be cleared or improved might encourage investment in housing but if local crime remained a problem such housing would remain unpopular and difficult to let. The 'total approach' understood the necessity for intervention in many aspects of urban policy at the same time.

The study was based in the Edge Hill area and focused on housing conditions, the quality of the physical environment, the effectiveness of local authority service delivery, and the impact of regional economic restructuring. Excluding the area management experiment, the team invested over £600,000 in local projects. The IAS represented the most detailed study yet of the problems of the inner areas. By bringing together an understanding of recent advances in urban theory with local empirical investigation a powerful report was produced. It provided detailed analyses of economic and demographic trends, the local housing market, housing need, local authority housing management, education, training and employment opportunities, the local physical environment, the delivery of local government services and relations between the local authority and the community. The study explored the difficulties of the unskilled, especially the young unskilled, in obtaining work at a time of rising unemployment. It analysed the pressures of bad housing and poverty on family life, and was one of the first reports recognise the issue of racial discrimination in Liverpool, particularly in local employment and housing markets. The report examined the reasons for low levels of industrial investment in the inner areas and offered one of the earliest detailed studies of vacant urban land – a matter of growing political concern in the mid seventies. One specific comment related to urban density:

> It must be accepted too that residential densities should probably continue to decline...One of the strongest components of urban stress is lack of space...It seems very unlikely therefore that reasonable living conditions will be possible in the inner areas without further reductions in living densities and an inevitable moving out of population (Wilson and Womersley, 1977, p202).

Amongst the most interesting parts of the study was the 'area management' experiment. The idea was to bring parts of the local authority administration closer to the people it was intended to serve, to foster a corporate approach to the needs of particular areas and achieve a fairer distribution of resources between areas (Wilson and Womersley, 1977, p20). The approach depended upon developing a detailed analysis of how much the local authority spent in the area, relative to other parts of the city. In other words: did the area get its fair share of resources? Did it benefit from positive discrimination? A small area management unit was established and led by an 'Area Executive' who reported to a district committee of local ward councillors. The experiment ran for about four years from 1975 to 1979. Whilst the concept was interesting, only limited success was achieved and area management failed to become part of mainstream policy. Proposals for area management were confronted by 'the strength of departmental viewpoints at all levels of the official hierarchy' and that 'part of the problem from the start was that there was no strong political backing for the experiment in Liverpool' (Wilson and Womersley, 1977, pp171-172). The analysis of local government spending on a small area basis proved exceedingly difficult and there were evident political tensions between the needs of the area and the priorities of the city as a whole. Nevertheless, Wilson and Womersley felt that the experiment was useful in showing how an area-based non-departmental focus on needs and issues could give fresh insights into the workings of the local authority. Furthermore, the policy reviews carried out under the guise of area management were influential in their effect on central and city-wide policies. But this is saying no more than in-depth local case studies are a useful mechanism for policy evaluation and development and has little to do with area management as such.

The IAS report concluded that although housing conditions had improved for many people, these gains were being put at risk by the under-funding and inefficient delivery of housing maintenance and repair. It was acknowledged that densities had been reduced, giving environmental benefits but this had to be balanced against the impacts of population loss on residual communities, services and the local economy. Urban deprivation was a serious problem in the inner areas and in pockets, a severe problem. It was argued that these problems stemmed in large measure from the persistence of divisions of status and income in society at large. Furthermore, the continuance of inner area problems reflected structural changes in the regional and wider economy. Central government was blamed for not understanding the extent of social and economic divisions within the city, whilst the local authority was criticised for the quality of service delivery and a lack of attention to the economic plight of the city (Wilson and Womersley, 1977, Ch 11).

A Model for Urban Regeneration Policy

The report recommended that the regeneration of the inner areas should be pursued through four interconnected programmes, which were seen as a necessary but not a sufficient basis for the regeneration of inner Liverpool. The fundamental requirement was for a total approach towards the inner areas that would build up a comprehensive understanding of its problems, and take concerted action for their solution (Wilson and Womersley, 1977, Ch12). Table 4.1 shows the content of the four interconnected programmes devised by the Liverpool Inner Areas Study. This was one of the first modern comprehensive models for urban regeneration policy.

Table 4.1   Liverpool Inner Areas Study: regeneration proposals

Promoting the economic development of Liverpool
> Prepare an economic development plan
> Promote industrial prospects amongst potential investors
> Increase the supply of serviced land and advance factories
> Prioritise large firms in the allocation of regional grants
> Encourage indigenous growth of small firms through support

Expanding opportunities for training
> Make regular assessments of manpower and training needs
> Reorganise and expanding training
> Set up integrated work/training programmes
> Improve the employment transfer scheme to help people leave

Improving access to housing for disadvantaged groups of people
> Regularly review of housing need
> Increase the supply of housing land
> Define new social needs priority in council housing
> Ease transfers between council tenancies within and beyond Liverpool

Channelling resources to areas of greatest social need
> Decentralise council housing management and tenant participation
> Develop community primary schools, pre-school and adult provision
> Improve take-up of income support and welfare benefits
> Allocate more resources to the voluntary sector

Source: Wilson and Womersley, 1977

Liverpool Inner City Partnership

Returning to power in 1974, the Labour government was coming to grips
with the emerging inner city problem. By 1975 it had the benefit of
feedback from the Community Development Projects and interim findings
from the Inner Area Studies. The Shelter Neighbourhood Action Project
had been completed and its director, Des McConaghy, had spent two years
as a special advisor to the Department of the Environment. The Secretary
of State for the Environment, Peter Shore, was also directly informed
through consultations with the major inner city local authorities. By the
following year the government were ready to synthesise all this analysis
into policy. In a landmark speech in Manchester on 17th September 1976
Peter Shore acknowledged the nature and severity of the problem:

> All our major cities have lost population over the last decade and a half.
> Since 1961 the inner area of the Manchester conurbation has lost 20 per cent
> of its population and that of Liverpool 40 per cent. What is more worrying is
> the unbalanced nature of the migration...leaving the inner areas with a
> disproportionate share of unskilled and semi-skilled workers, of
> unemployment, of one-parent families, of concentrations of immigrant
> communities and overcrowded and inadequate housing. Though there has
> been a growth of office jobs in the centres of most of these cities this has not
> compensated for the extremely rapid decline in manufacturing industry in the
> inner urban areas. Manufacturing employment in Manchester declined by 20
> per cent and Liverpool by 19 per cent between 1966 and 1971 (Shore, 1976).

Shortly afterwards the government published their 1977 White Paper
'Policy for the Inner Cities'. Within the overall context of the post-1973
recession, there was a questioning of the value of an area-based approach to
regeneration and recognition of the link between urban deprivation and the
operation of the wider economy. The White Paper also acknowledged the
social and fiscal problems facing many urban authorities and the need to
review the roles and responsibilities of government departments with
regard to these issues.

Through the Inner Urban Areas Act 1978 and other related policy
initiatives the government sought to establish a framework for urban
regeneration. It was accepted that local authorities were to take the lead in
spearheading action in the inner urban areas. Nevertheless, lack of co-
ordination between central and local government was thought to be a major
problem. In consequence the Act established 'Inner Area Partnerships' in
the seven most acutely deprived urban areas: Birmingham, Hackney,
Islington, Lambeth, Liverpool, Manchester/Salford, and Newcastle

/Gateshead. These areas would benefit from a partnership between central and local government to achieve economic regeneration through co-ordinated inter-corporate action. To facilitate this each area would have a 'partnership committee' chaired by a government minister. These areas would also receive the top priority for funding.

Responding to the rising pressures on local authority services in the inner urban areas the formula for distributing the Rate Support Grant (the mechanism for redistributing tax income from richer to poorer areas) was to be adjusted to redirect funds away from the shires towards urban areas. The Urban Programme was revitalised and transferred from the Home Office to the Department of the Environment, which became the lead department for urban regeneration. The development programmes of new towns and expanded towns were to be progressively wound down in recognition of the fact that out-migration was now damaging the economy and social life of the inner areas. There was to be more emphasis on the reclamation of vacant and derelict urban land. Area improvement was to be preferred to slum clearance and housing associations were to play a growing part in this improvement process.

It was in this context that the Liverpool Inner City Partnership was established. A joint working party of officers was also set up to ensure the more co-ordinated approach that was required. The inner city partnership included most of the city within Queens Drive together with the outlying area of Garston. The overriding objective was to be the economic, social and environmental regeneration of inner Liverpool. According to the Partnership this meant reversing population and economic decline through policies to make the inner city as desirable and attractive as possible. It was also thought important to concentrate action on selected areas of the inner city so that the maximum impact could be obtained (Liverpool Inner City Partnership Committee, 1978, p29).

One of the first tasks of the Partnership Committee was to agree a spending programme for the initial 1979-82 period. This was based upon a detailed analysis of the local economy and the needs of the area. In all a programme of some £48 million of public expenditure was agreed. This represented an increase of approximately 30 per cent over previous levels of expenditure (Liverpool Inner City Partnership Committee, 1978, p10).

But much of the Labour government's new policy for the inner cities would never be implemented. May 1979 brought the election of Margaret Thatcher's Conservative Government, with some very different ideas about urban regeneration from their Labour predecessors.

Lessons from a Decade of Urban Investigation

Thus between 1968 and 1978 there was a decade of experimentation and development in urban policy. The era began with a simplistic analysis that suggested the causes of urban deprivation could be explained by notions of a 'culture of poverty' and a 'cycle of deprivation'. Gradually other explanations were added including the idea of failures in the management and delivery of services by central and local government, the inadequacy of the resources available to the inner cities, and the impact of structural changes in the national economy, beyond the control of local people.

The idea of action-research as a mechanism to develop urban policy was extensively used by many studies around this time, although it was by making connections between local areas and by comparing local and national trends that researchers were able to establish the true structural nature of the underlying causes of urban deprivation. Such problems could not be solved without radical changes to existing urban policies, institutional structures and most important of all, substantial increases in the amount of public funding for the provision of services in the inner cities and to finance the process of regeneration. It was clear that the response had to be holistic (the total approach) and comprehensive in tackling all the causes of urban deprivation at the same time. With this came recognition of the need for co-ordination between the various agencies and levels of government (joined-up thinking).

By the end of the decade the government had taken on board many of these conclusions. Through the Inner Urban Areas Act 1978 and associated policy changes, a set of measures and delivery mechanisms had been devised, including the nascent Liverpool Inner City Partnership. The idea of partnership was permitted by the closeness of view and sharing of objectives that existed between central and local government at that time. Whether these policies would have successful in tackling inner city deprivation is uncertain. Clearly the approach was not politically robust for it was in large measure abandoned by the incoming Conservative government after 1979.

Even from the beginning of the decade urban planners held the view that if spatially concentrated pockets of urban deprivation were to be eradicated then the whole urban system in its regional setting needed to be understood and planned in a comprehensive manner. Preliminary work on a structure plan for Liverpool had commenced in 1971 but was quickly abandoned after the passing of the Local Government Act 1972. With the reorganisation of local government, strategic planning powers were removed from the City Council and passed to the new Merseyside County

Council on the assumption that it could take a broader view of the city and its region. In consequence a 'Merseyside Structure Plan Team' was established in 1972 to begin the process of preparing a strategic plan for the future development of the whole Merseyside conurbation on a rational comprehensive basis. Tackling urban deprivation was to become one of its key objectives.

5 Strategic Planning for the Conurbation

Under the provisions of the Town and Country Planning Act 1968 unitary planning authorities were to be responsible for strategic policy, tactical planning and much of the implementation of plans. The Act extended the scope of planning to include not just control over development and the use of land but also transport policy, conservation and improvement of the physical environment. In the larger urban areas county borough councils were to prepare urban structure plans that would provide a framework for their own local plans. The same authority would then provide much of the infrastructure for development, together with schools, community facilities, council housing and the control of private development. Elsewhere county councils would prepare county structure plans providing a similar framework for local plans and development control. The county council would provide schools and much of the infrastructure whilst urban and rural district councils would provide council housing and many community facilities.

Despite the strong commitment to planning inherent in the passing of the 1968 Act, within a dozen years the whole system had been undermined to such an extent that structure plans were held to be of little value and some commentators even questioned the very basis of the planning system. This process had begun even as the Act was reaching the statute book. Although the Act contained provisions for local planning authorities to include policies for the conservation of the built environment within local plans, the Civic Amenities Act 1967 already allowed local authorities to designate 'conservation areas' within which a stricter than normal planning regime could be exercised. One of the main attractions of 'conservation areas' (CA) under the 1967 Act was speed of implementation. Technically a local plan could not be prepared until the structure plan had received ministerial approval – a process that could, and did, take years. On the other hand a 'conservation area' could be brought into force through a decision by the local authority acting within its own powers.

Similarly, the 1968 Act contained provisions to allow local planning authorities, after approval of the structure plan, to prepare detailed 'action area plans' for areas where rapid development or change was anticipated. These might include, for example, districts that were to benefit from housing renovation and area improvement. Again another act of parliament offered an alternative route to the same end. The Housing Act 1969 permitted local authorities to designate a 'general improvement area' (GIA) within which they could plan and co-ordinate area improvement policies. Not only did such a concept replicate what could have been included within an action area plan but the GIA could be designated by a decision of the local authority acting within its own powers and furthermore attract central government funding to support environmental improvement works. Also passed in 1969, the new Transport Act established Passenger Transport Authorities in the major conurbations, including Merseyside, with responsibility to operate *and plan* all local public transport.

But the biggest threat to planning came with the re-organisation of local government under the provisions of the Conservative government's Local Government Act 1972. This radical reform of the structure of local government established a two-tier system of county councils and district councils across the whole of England and Wales. The preparation of development plans was to be split with county councils taking responsibility for structure plans and some local plans, and district councils preparing most local plans and handling most development control. According to Ward:

> This severely compromised the integrity of the whole 1968 Act system which had rested on the assumption of unitary authorities. It made it impractical to assume that local plans, produced by districts, could simply be the tactical elaboration of the counties' strategic structure plans, since there was a greater likelihood of real conflicts between plans produced by two different authorities (Ward, 1994, p140).

One of the strongest arguments in favour of structure plans was their ability to plan the all-important relationship between land use and transportation. The 1968 Act facilitated, for the first time, the strategic planning of land use developments and the provision of transport infrastructure in the same document. So keen was the government to bring together land use planning and transport policy that in 1971 the Ministry of Housing and Local Government and the Ministry of Transport were merged into one new ministry: the Department of the Environment. However, within little more than a year the government undermined this linkage with a statutory obligation on the new county councils to produce annual 'transport policies and programmes' (TPP). Despite being expected to *have regard* to structure plans, the TPP, being annual and linked to the allocation of funding, quickly became the principle means of expressing transport policy and became increasingly detached from the land use planning process. Only two years after the introduction of the TPP the government similarly required the new district councils to prepare an annual 'housing investment strategy and programme' (HIP). This too contained a statement that policy, like the TPP, should *have regard* to the provisions of the development plan. It was not surprising that the HIP, rather than a local plan, quickly became the more important and more easily updated mechanism for expressing a local authority's policies for housing development. Table 5.1 indicates the changes in planning responsibilities in Liverpool that occurred between 1968 and 1974.

Table 5.1   **The planning system in Liverpool under the Town and Country Planning Act, 1968 and subsequent arrangements**

1968 - 1974
Structure Plan   Liverpool City Council (City Planning Department)
Local Plans   Liverpool City Council (City Planning Department)

After 1974
Structure Plan   Merseyside County Council (County Planning Department)
Some Local   Merseyside County Council (County Planning Department)
Plans
Most Local   Liverpool City Council (City Planning Department)
Plans
Transport Policy   Merseyside County Council (Joint Transportation Unit)
and Programmes
(TPP)
Housing   Liverpool City Council (Housing Department)
Investment
Strategy and
Programme
(HIP)

Source: author

Thus only a few years after the passing of the Town and Country Planning Act 1968 the development plan system was in deep trouble. Even by 1978 many counties, including Merseyside, still did not have an approved structure plan. It was axiomatic that there were no approved local plans or action area plans in such areas. In Liverpool the only valid statutory development plan had been approved back in 1958: little wonder that planning was coming to be held in low regard. For this reason, in Liverpool as much as in any other city, the development plan process was increasingly being by-passed. The whole planning process was becoming fragmented as alternative means (CA, GIA, TPP, HIP) were being used to achieve policy aims that the new development plan system had been intended to formulate and co-ordinate. The dream of a rational comprehensive planning system was collapsing.

It is therefore surprising that despite all the frustrations and fragmentation of urban planning the nineteen seventies can still be seen as a decade in which there were important developments in strategic planning. It is particularly surprising that two of these plans were of relevance to the Liverpool conurbation. The first of these was a regional plan: the Strategic Plan for the North West (North West Joint Planning Team, 1974) and the second was the Merseyside Structure Plan (Merseyside County Council, 1979). Between them they represented ambitious and heroic attempts at rational comprehensive planning at both the regional and sub-regional scales.

Strategic Plan for the North West

The Strategic Plan for the North West (SPNW) was one of the last in a series of regional plans commissioned by the Government for different parts of the country. Work started in 1971 and a team of officers, mainly seconded from Cheshire and Lancashire County Councils and the Department of the Environment, took only two years to produce a regional plan extending to nearly 300 pages, together with a number of technical reports.

The purpose of the plan was to provide a framework for government decisions and policy-making in the region as well as strategic guidance to local planning authorities preparing their development plans. It was not seen as a rigid master plan but the basis for a 'continuous process of regional planning'. SPNW was published at the very moment that local government reorganisation was taking place. The timing of the report is also interesting because it was being prepared at the nascence of green politics (Friends of the Earth, 1972), and after the publication of one of the earliest reports to link continued economic growth with global environmental change (Meadows, 1972). Significantly the first substantial chapter of SPNW was entitled 'How Much Growth?' and included the following paragraph:

> The principal proposition put forward by the so-called 'environmentalists' is
> that the world's resources are finite: if growth in both population and
> economic terms is allowed to continue indefinitely the eventual result could

be the disappearance of an acceptable human environment (North West Joint Planning Team, 1974, p23).

This was followed by a debate on the merits of pro and anti growth strategies: one of the first times such a debate occurred within a regional plan. Nevertheless the report's working proposition was that there should be limited population growth in accordance with current trends and continued economic growth in order to achieve better standards of living for all (North West Joint Planning Team, 1974, p39). The plan then confined itself to consideration of where and under what constraints such growth should be accommodated. Further chapters dealt with 'opportunities and needs', 'resources', 'public expenditure priorities', 'the urban environment', 'employment policy for the region', 'regional transportation policy', 'regional open land policy', 'the physical pattern of development', 'the meaning of the strategy', and 'implementation and continuous planning'.

The North West was the worst region in the country in terms of certain environmental indicators: river pollution, air pollution and the incidence of derelict land. Social indicators such as infant mortality, the availability of doctors and pupil-teacher ratios also painted a bleak picture of the region. Nevertheless, the report recommended taking advantage of the 'enormous capital stock' of the region, suggesting that even deficiencies could be turned to good account, for example through the re-use of derelict land. The proposed strategy for the spatial distribution of physical development in the region was to concentrate investment in the Mersey Belt (essentially the conurbations of Merseyside and Greater Manchester and the area between them). The reasoning behind this strategy was that the aims of the plan were best met by a pattern of development that:

> emphasises 'concentration' within the Mersey Belt, especially in mid-south Lancashire and along corridors served by public transport; restrictions on the release of open land in North Cheshire; lower rate of loss of people and jobs from the metropolitan counties; Central Lancashire New Town as a major long-term growth area; and green belts to shape and retain development accordingly (North West Joint Planning Team, 1974, p19).

The need to improve living and working conditions in towns and cities was regarded as the predominant issue facing the region. These problems were concentrated in the Mersey Belt (including Liverpool), North East Lancashire and the salt towns of Cheshire. It was recommended that co-ordinated action by local authorities supported by additional central government funding was the best way forward. Unemployment was an emerging issue. The basic causes of the region's poor employment record were identified as too many declining industries and too few growth industries, i.e. a structural problem. The plan proposed a programme of research on industrial performance, actions to promote greater industrial efficiency; a labour subsidy to maintain a balance between capital and labour incentives, a review of office relocation incentives, improved training provision, and measures to assist the mobility of labour.

Considerable attention was devoted to the inadequacies and shortcomings of public sector funding in the region and recommended improvements in the rate support grant and increases in certain specific grants. A chapter on public expenditure focused on the costs and priorities for investment, although only the treatment of environmental pollution was worked out in detail. It was suggested that the most cost-effective programmes were for the reduction of air pollution (from industrial plants) and the reclamation of derelict land. Further, expenditure on reducing river pollution was to be selective in favour of rivers that supplied drinking water and rivers forming part of recreational systems, including improvement of the Mersey and its estuary.

In order to protect and optimise the use of open land it was argued that all such land in the region should be committed to uses designed to fit in with the broad physical pattern of proposed development. Policies included the creation of regional parks, opening up middle-distance footpaths, formulation and adoption of green-belts and provision of additional outdoor recreational facilities accessible by public transport. With regard to transport policy the plan acknowledged the contribution of two recently completed land-use-transportation studies in framing its proposals for the conurbation areas: MALTS, the Merseyside Area Land-Use Transportation Study; and SELNEC (South East Lancashire and North East Cheshire). Proposals included the upgrading and electrification of selected railway lines in the region and the building of a number of inter-urban roads. But even so it would be necessary to restrain car usage:

> If full value is to be obtained from improved public transport, restrictions on parking and use of road space must be applied in order to curtail car usage (North West Joint Planning Team, 1974, p12).

SPNW made a number of specific references to Liverpool itself. It was anticipated that over the twenty years from 1971 to 1991 the population of the city would decrease from 610,000 to between 520,000 and 540,000 (a drop of between 11.5 per cent and 14.8 per cent) compared with a regional expectation of 3.9 per cent population growth (Tables 2.8 and 10.2). It was recognised that Liverpool had the largest number, though not the highest percentage, of dwellings needing treatment of any local authority in the region except for Manchester. The strategy of concentrating growth in the Mersey Belt required powerful restrictions on expansion in Wirral, Sefton and the Ormskirk plain but corridors for new development (even at the expense of some green belt land) between Kirkby, Skelmersdale and Wigan, and between Huyton, Prescot and St Helens. In support of this development it was anticipated that by 1991 suburban passenger rail services between Liverpool and Wigan, St Helens and Warrington would be upgraded and electrified and that the outer rail loop and Edge Hill Spur would be completed. Extension of the M62 motorway as far as the Liverpool Inner Ring Road and the M57 south to Widnes were also proposed.

With its strong emphasis on the efficient use of land, improving the physical environment and amenity of urban areas and the protection of open land, this was a plan prepared in the ideological tradition of post-war British planning. However, in addition to these classic concerns the plan was innovative in examining the fiscal problems of local authorities and in giving considerable attention to the mechanisms and agencies through which the plan was to be implemented, as well as arrangements for its monitoring and review. Also new was the regard being paid to the emerging environmental agenda. The overall strategy of urban concentration, the reuse of existing resources (e.g. derelict land) and maximising the potential of the existing capital stock, was compatible with a 'sustainable' approach to development. There was a strong emphasis on the greater use of public transport with encouragement for further investment in public transport systems and a recommendation that major developments be concentrated in locations that would be easily accessible by public transport.

In subsequent years only parts of this regional planning strategy have been realised. Local authorities have developed and implemented strong policies for the protection of open land, though with some notable breaches. Much has been done to improve the physical environment of urban areas. Large amounts of derelict land have been reclaimed and returned to beneficial use. Air pollution from industrial sources and river pollution have been reduced. Less success has been achieved in relation to the economy where unemployment rates, especially within Liverpool, have remained high relative to the national average. There has been a similar failure in transportation policy with a conspicuous inability to invest adequately and appropriately in the system. Few of the public transport investment schemes proposed for Liverpool have come to fruition. But perhaps of most concern has been the failure to respond to the fiscal problems of local authorities, especially in Liverpool.

Merseyside Structure Plan Team

The preparation of the Merseyside Structure Plan was intertwined with the emergence of the Labour government's own policy for the inner cities. Liverpool City Council had begun preliminary work on an urban structure plan back in 1971 but work effectively stopped when the reorganisation proposals of the Local Government Bill became known. Instead Liverpool joined forces with neighbouring authorities to create the 'Merseyside Structure Plan Team' led by Peter Wood, who later became Deputy and then County Planning Officer for the new county. This team comprised planners seconded from Liverpool, Birkenhead, Wallasey and Southport County Borough Councils, and Lancashire and Cheshire County Councils. Work commenced in 1972 with the aim of producing a draft structure plan in time for the creation of the new county in April 1974. Guided by a Steering Group of senior politicians from the constituent authorities, the team set about the 'technical' process of structure plan production with some vigour. It was anticipated that the process would go through three iterations or cycles: a first 'quick and dirty' cycle would produce a rough draft plan within a couple of months that would clarify aims, objectives and major issues; a second much longer cycle of 12-18 months would provide considered policies supported by reasoned justification and evidence; and a third short cycle would refine the plan after consultation.

One of the earliest publications from the team was 'Merseyside: A Review' (Merseyside Planning Officers Group on Structure Planning, 1972). This report noted that the population of Merseyside (including Liverpool) had been falling in recent years. There had also been a drop in employment accompanied by significant structural changes. There remained in the county a large stock of obsolete housing. Changing shopping habits, redevelopment and rising car ownership were all having an impact on the pattern of shopping centres. This burden of decline impacted particularly on the older, inner areas that also suffered from poverty and a multitude of social problems. The environment of many of these inner areas was also often poor. Traffic congestion was a major problem and increasing road traffic was one of the major causes of environmental degredation. The countryside of Merseyside was vulnerable to a variety of urban pressures (Merseyside Planning Officers Group on Structure Planning, 1972, p2).

In this context the principle aims proposed for the structure plan were to:

- cure the persistently high level of male unemployment and give Merseyside a healthy economy
- remove the remaining inadequacies of Merseyside's housing
- remove the detrimental conditions which adversely affect health, educational attainment and social well-being
- provide for the leisure needs of the population
- correct the mis-match between shopping provision and people's needs
- improve accessibility throughout Merseyside and reduce the noise and danger from traffic
- improve the townscape and landscape of Merseyside and conserve its best features
- replace or modify inadequate methods of waste disposal and other activities which pollute the environment

This was a markedly different set of priorities from the IPPS of only eight years earlier. That document, admittedly working at a slightly different scale, had put most emphasis on creating an efficient urban form and transportation system, whereas in this document urban form was not seen as a major issue and transport policy was relegated to a concern about traffic. Even housing was no longer seen in terms of critical shortage as it had been in 1965. In contrast employment and social conditions were now of major concern to planners. The inclusion of waste disposal and environmental pollution reflected a new understanding of environmental capacity and the limits to growth. These concerns were similar to those considered by SPNW and reflected a changing political agenda and the influence of social scientists and geographers on the planning profession.

In January 1973 the team produced an intriguing report entitled: 'Merseyside 1986: if present trends and policies continue' (Merseyside Structure Plan Team, 1973). It was anticipated that the population of Liverpool would fall from the 1971 figure of just over 600,000 to between 500,000 and 565,000 by 1986. Overall in the county:

More people would own a house and car but a substantial section of the population would still be relatively poor, employment would be static and 100,000 families would be in houses about 100 years old. Opportunities for leisure and the quality of the physical environment would not be particularly good; an inadequate road system and a good, but underused and expensive public transport system could exist side by side. At the same time, the standard of service expected by people would be making increasing demands on public services (Merseyside Structure Plan Team, 1973, p2).

In fact their key projections proved optimistic. By 1986 the population of Liverpool actually stood at around 480,000 and employment in the city had fallen by around 30 per cent. Slum clearance rates fell dramatically after the Housing Act 1974, consequently many of the oldest dwellings in 1971 were still standing in 1986. Most of the other predictions about the state of the environment, transport provision and the pressure on public services proved disturbingly accurate.

The team regarded public participation as 'one of the most important aspects of work on this Plan' and throughout 1972 undertook a programme of three surveys. The first survey sought the opinions of local authorities and organisations on the aforementioned 'Merseyside: A Review'. The second was an interview survey with a representative sample of Merseyside householders to find out their views and attitudes, and the third was a postal survey of aldermen and councillors on problems in their wards. In addition twelve expert consultative groups considered such topics as Commerce and Industry, Housing, etc. (Merseyside County Council, 1974).

The results of this consultation process were interesting. The household interview survey had been undertaken by a professional research institute (Social and Community Planning Research). The most *chronic* problems to emerge were 'vandalism, dirt and smells'. Other serious problems included the desire to move, the poor physical condition of dwellings, aspects of the environment such as the general appearance of an area, the expense of public transport, the lack of sports and play facilities, and unemployment. The survey also identified a further series of issues that were less widespread but which were *acute* problems for those experiencing them. The worst of these were unemployment and job insecurity. These were followed by poor services for the elderly and the handicapped, insecurity of tenure and poor household facilities. This list was rather different from the aims for the draft structure plan identified in the 1972 review. Perhaps predictably the survey of aldermen and councillors identified housing and the local environment as serious problems, particularly in the inner area wards, with transport also seen as an issue, especially in outlying areas (Merseyside County Council, 1974).

The aim of producing a draft structure plan by April 1974 proved impossible to achieve for two reasons. Firstly, the impending reorganisation of local government led to major staff changes throughout the latter half of 1973 and the first quarter of 1974 which destabilized and delayed the plan production process. Secondly, the incoming county councillors made it clear that they, rather than any team of officers, would determine the nature of any strategy, thus emphasising that the structure plan was a political rather than a technical document. This was not least because it was seen as one of the key policy tools available to the county council and one of the few mechanisms that it could use in negotiations with other agencies such as district councils, government departments and the embryonic regional health and water authorities.

The new county planning department, under the direction of Audrey Lees, was an interesting and innovative organisation. It was the smallest planning department in any of the new metropolitan counties, employing only a modest establishment of highly qualified professional staff relying on consultants to provide specialist advice and undertake time consuming survey and analytic work as necessary. This novel approach of 'privatising' or 'sub-contracting' planning work was said to increase the flexibility in responding to changing issues. The department was sub-divided into three sections including structure plan preparation, development control and the environmental protection unit. This last section was a new phenomenon within a county council and brought together experts in ecology, natural resource management, derelict land reclamation and conservation. It placed Merseyside County Council amongst the leading authorities in local environmental policy development in Britain at that time.

The delay caused to the preparation of the Merseyside Structure Plan by the reorganisation of local government was substantial. Even by the middle of 1975 there was little evidence of progress and an urgent need to fill the strategic policy vacuum. After all, if the County Council could not come up with a strategic plan for the county, what was it for?

'Stage One Report'

To fill this policy gap, a team of planners (led by Tony Struthers, who would later become President of the Royal Town Planning Institute) prepared a brief statement of strategic policy that became known as the 'Stage One Report' (SOR) (Merseyside County Council, 1975). It was adopted in September 1975 as an interim statement of planning policy. This report rehearsed the impact of past policies on Merseyside, the county's assets and influences for change. It was noted that despite the problems of depopulation and disinvestment emerging in the inner areas, policies were still encouraging decentralisation and overspill. There was under-use of social and physical capital and a need for investment in the inner areas whilst at the same time outward growth was increasing pressures on the countryside and requiring additional investment in infrastructure and service provision. Such a situation was inefficient in the use of economic resources, environmentally destructive and socially damaging. At the same time the inner areas contained physical and social capital that, if properly managed, could be utilised to support economic growth and regeneration. (Merseyside County Council, 1975, pp 4-6). The strategy proposed to:

> Concentrate investment and development within the urban county and particularly in those areas with the most acute problems, enhancing the environment and encouraging housing and economic expansion on derelict and disused sites. It would restrict development on the edge of the built-up areas to a minimum. There would be a reciprocal effort to enhance and conserve the natural features of the county's open land and its agriculture while ensuring that its capacity to meet the county's needs for leisure, recreation and informal education is exploited (Merseyside County Council, 1975, p8).

The strategy was based on the idea that by restricting peripheral growth development could be forced back into the inner areas. Whether or not this could be achieved was unknown. In the years leading up to 1975 industrial investment in the inner areas had been minimal and private housebuilding virtually non-existent. Private investors favoured green-field sites for ease of development and profitability. Even government agencies encouraged the suburbanisation process: 'new towns' were still being developed at Runcorn, Skelmersdale and Warrington specifically to accommodate Liverpool overspill. The English Industrial Estates Corporation concentrated much of its investment in advance factories and serviced industrial land in suburban locations. Motorways and other major roadbuilding programmes rarely penetrated existing urban areas because of cost.

All this would have to change. Land use planning would have to restrict peripheral development. A strong green belt policy would need to be enforced. The new towns programme would have to be wound down. Government investment in transport and industry would have to be redirected. In the inner areas ways would have to be found to encourage private housing investment and the return of industry. Derelict land and buildings would have to be brought back into beneficial use.

The Stage One Report represented a major shift in planning policy on Merseyside. Although no more than a brief statement of an approach to the problems of the conurbation it was one of the first statements of modern urban regeneration policy to emerge from an English local authority. Yet it would take another twenty years of debate, experimentation and conflict before regeneration would mature into an accepted part of mainstream government policy and even then everything about the process, from its strategic aims to the tactics of implementation, would remain highly controversial.

Merseyside Structure Plan

The Merseyside Structure Plan itself was not published for another four years. The introduction to the plan described it as:

> a guide for Merseyside for the 1980s. It points to the opportunities for investment, the assets which must be conserved, the problems facing the people of the County and the ways they should be tackled. It offers a hard-headed assessment of the money likely to be available and what it is possible to achieve...The plan concentrates on finding practical solutions to...problems and describes some of the ways Merseyside can attract investment (Merseyside County Council, 1979, p1).

Back in 1965 in the IPPS Alderman Sefton had expressed huge ambitions for the modernisation of the city with the full expectation that they would be delivered through the realisation of the plan. Here in 1979 the authors offered only a guide to what might be possible.

One of the reasons for the delay in the production of the structure plan had been that despite the government's broad acceptance of the emerging strategy outlined in the Stage One Report, it still required the County Council to go through a process of testing alternative strategies in order to provide a reasoned justification for their preference. (Merseyside County Council, 1979, p7). Thus three alternative strategies had to be developed and tested. These were: urban regeneration; managed dispersal; passive decline.

Unsurprisingly, the evidence showed that a strategy of passive decline would lead to a dwindling population bearing an increasing financial burden for services whilst suffering a worsening quality of life. The inner areas would enter a spiral of decline. Managed dispersal (the continuation of overspill) was unnecessary and would not help solve the problems of the most needy. Only a strategy of urban regeneration would make use of wasted inner city assets and offer the chance to protect the countryside from development.

The chosen strategy, urban regeneration, was described in terms of its implementation and impact in four different zones: older urban areas; outer council estates; suburban areas; country areas. In the older urban areas the main aim of the strategy was to restore confidence in order to attract private investment. The main planning polices and proposals, insofar as they affected Liverpool, are illustrated in Figure 5.1.

The County Council would offer information, advice and financial aid to businesses and would seek to remove legal, planning and other obstacles. This was one of the first times that 'planning' was seen by the planning system itself as an *obstacle* to development: a reflection of successful lobbying by the property sector. The emphasis was on encouraging the development or redevelopment of underused or unused land and buildings. The nature of that development, historically one of the key concerns of planning, was evidently of lesser importance. Through the County Council's own economic development office and other agencies, local firms would be encouraged to expand and new employers encouraged to move to Merseyside. Efforts would be made to find developers for vacant and derelict land and in particular, private housebuilders would be encouraged to develop vacant sites in older urban areas with houses for sale (Merseyside County Council, 1979, p8).

Large-scale slum clearance was no longer necessary. Improvement was the preferred option for treating obsolete and substandard housing. A plea was made to the government to regularly review the level of improvement grants and the criteria for designating general improvement areas and housing action areas. This too was something new. The plan was being used as an *advocacy* document to encourage changes in government policy.

The only significant reference to other aspects of the physical environment of the inner areas was a recognition that such areas are often short of good quality parks and sports facilities. Although transport policies were discussed in a later detailed chapter, they were not mentioned in relation to strategy for the older urban areas.

Outer council estates were for the first time in a major plan seen as part of the problem rather than part of the solution. Many estates were becoming characterised by high unemployment and a dearth of local employment opportunities. The proposed solutions included making land available for nearby industrial development, encouraging local office developments, training programmes, and better public transport to improve access to employment, shopping and social facilities. Housing on some of these estates was no longer of a standard that was acceptable to tenants and there was need for improved maintenance, renovation or (in extreme cases)

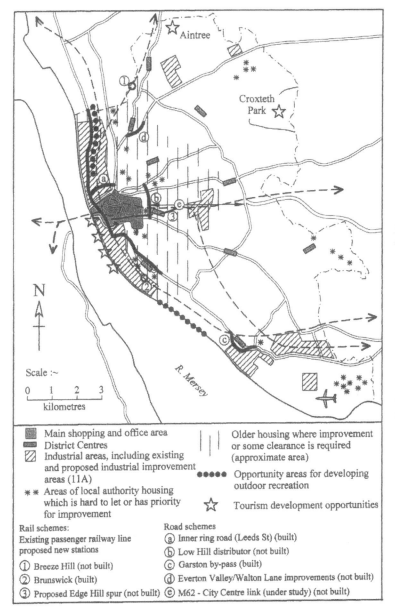

Scale :~

0   1   2   3
kilometres

	Main shopping and office area
	District Centres
	Industrial areas, including existing and proposed industrial improvement areas (11A)
* *	Areas of local authority housing which is hard to let or has priority for improvement

Older housing where improvement or some clearance is required (approximate area)

●●●●● Opportunity areas for developing outdoor recreation

☆ Tourism development opportunities

Rail schemes:
Existing passenger railway line
proposed new stations

① Breeze Hill (not built)
② Brunswick (built)
③ Proposed Edge Hill spur (not built)

Road schemes
ⓐ Inner ring road (Leeds St) (built)
ⓑ Low Hill distributor (not built)
ⓒ Garston by-pass (built)
ⓓ Everton Valley/Walton Lane improvements (not built)
ⓔ M62 - City Centre link (under study) (not built)

Figure 5.1   **Main policies and proposals from the Merseyside Structure Plan, 1980 and the Statement of Current Planning Policies and Development Proposals, 1983**

Source   Adapted from Merseyside County Council, 1980 and Liverpool City Council, 1983 demolition.   It was also suggested that a greater social mix could be encouraged on large council estates by allowing private housebuilding on cleared and vacant sites.

Little change was expected in other suburban areas. The approach was to conserve and enhance the local environment, whilst permitting housing and industrial developments on land that had already been earmarked for such purposes.   In the country areas the County Council envisaged designating a green belt to limit development on open land and to protect high quality agricultural land. Areas of importance for nature conservation, sites of scientific interest and heritage landscapes were to be protected whilst damaged landscapes were to be restored.   Countryside recreation was to be encouraged.

Despite the emphasis on urban regeneration in the overall strategy, the structure plan contained policies that dealt with a wider range of issues than previous plans for the city.   In addition to policies for the economy, housing and transport, the plan also considered the urban environment, natural resources, open land, the green belt, recreation, tourism and shopping.

With regard to the economy, it was argued that without new initiatives the gap between the numbers of jobs and people of working age would widen.   Across the county it was anticipated that this gap, which stood at over 100,000 jobs in 1976 could widen to 150,000 by 1986.   It was acknowledged that many of the solutions lay beyond the scope of the structure plan, whose policies were limited to the provision of land for industrial development. In addition to a spread of industrial land across the conurbation, provision was to include at least one strategic site capable of accommodating a very large industrial project, such as a motor vehicle manufacturing plant.

It was now accepted that the re-location of 'non-conforming industrial uses' away from residential areas was no longer practicable nor desirable. Local authorities would in future normally grant permission for the extension or modernisation of existing industrial plants, subject to local planning conditions.   This was a change in policy.   Furthermore, small industrial sites, serviced industrial land and advance factories were to be made available in areas where local job opportunities were restricted and travel to work was difficult. The designation of industrial and commercial improvement areas would facilitate improvements to the local environment, security, accessibility and site assembly.

Major office developments would only be permitted in town and city centres such as Liverpool, Birkenhead, Bootle, Huyton, Kirkby, St Helens and Southport. Small office developments would be permitted in other urban centres to serve local needs. Nevertheless, almost in contradiction to this centralising policy, it was stated that:

> To meet the needs of employers with special requirements, which cannot be met within the main office centres, sites will be assembled, serviced, landscaped and promoted as prestige locations for out of centre office/research and development facilities at Knowsley (near the junction of the M57/A57) and at Upton, Wirral (Merseyside County Council, 1979, p32).

This was one of the first explicit indications that economic development was a more important policy aim than an efficient transport system or environmental protection.

Despite a discussion of the pros and cons of housing clearance or improvement as methods for treating obsolete housing, when it came to policy, the plan made no distinction between the two alternatives, arguing that the amount of clearance would depend largely on the future rate of improvement. Within Liverpool itself it was suggested that between 25,000 and 38,000 privately owned dwellings should be improved or cleared between 1977/78 and 1985/86: hardly a precise figure. For the first time in a plan for Liverpool it was recognised that housing associations would play an important part in the housing regeneration strategy. Some 31 per cent of Merseyside's council dwellings were also in an unsatisfactory condition. Within Liverpool it was proposed that between 16,000 and 30,300 local authority owned dwellings should be improved or cleared between 1977/78 and 1985/86: again the figures were so imprecise as to be almost meaningless.

Land for housing development was to be provided in line with house-building trends. Across the county this meant the provision of 1031 hectares of land for development between 1979 and 1986, of which 397 hectares would be in Liverpool. In contrast with the figures for obsolete housing such precision was remarkable! In allocating land for housing development the district councils were expected to minimise the need for new services (utilities), encourage the development of unused and derelict sites, minimise the use of good quality agricultural land, and encourage the development of land with good access to public transport services.

One of the most important policies for the urban environment was the reclamation of vacant and derelict land. In the short term it was proposed that vacant sites should be landscaped so as to enhance the appearance of the area. In the longer term the intention was to bring such land back into use. The preference was that vacant and derelict land should be re-used for industry and housing. In practice industrial re-use proved difficult to achieve and a high proportion was redeveloped as open space, housing or other uses. In fact, between 1981 and 1988, 50 per cent of vacant sites in Liverpool were redeveloped for housing, 14 per cent for retail or distribution uses, 32 per cent for other uses and only 4 per cent for manufacturing industry (Couch and Fowles, 1992, table 7.6).

Another important policy for the urban environment was the restoration and extension of existing parklands. It was also felt that in many parts of the conurbation the urban landscape could be greatly enhanced by a programme of tree planting, although at this point in time the justification for tree planning was aesthetic rather than ecological. District councils were encouraged to conserve areas and buildings of architectural or historic interest. Within residential areas the local environment was to be improved, where necessary, by the removal of eyesores and dangerous structures, landscaping of undeveloped sites and resisting the development of environmentally intrusive industry or commerce. Smoke control programmes were to be extended to those areas where they still had not been implemented. The most serious backlog concerned older council housing. More than 90 per cent of the inner area council estates were still not smokeless zones in 1979, 23 years after the 1956 Clean Air Act (Merseyside County Council, 1979, p59). New developments would not be permitted where the proximity of existing industry would pose a risk to health. At the same time local authorities would seek reductions in environmental pollution from industrial sources. Selected locations would be defined as suitable locations for 'special' (i.e. polluting) industrial development. There were also policies for the treatment and disposal of waste. Materials and energy were to be recovered from waste. This recycling of materials would occur at waste treatment plants on the basis that it would reduce the need for landfill sites – an emerging problem.

The plan contained a number of policies to protect natural resources and open land. The most important policy directly affecting Liverpool itself was the cleaning up of the Mersey. New interceptor sewers were being constructed along the shore to take sewage to new treatment works before disposal into the estuary. Virtually all the open land surrounding the conurbation was to be designated a 'green belt', effectively prohibiting any substantial development of the remaining open land in the county. This policy was implemented through the approval of the Merseyside Green Belt Local Plan in 1983.

Given the recent decline in population and employment, particularly in the inner areas, it was recognised that the emphasis of transport policy would be on resolving existing problems rather than investing in infrastructure to support new urban development. The main issues were seen to be the need to find an acceptable balance between the level of public passenger transport services, fares and the amount of subsidy; minimising conflict between traffic and the environment; and preventing the environmental and physical deterioration of some areas. Thus policy was to:

- give priority to maintaining and improving existing roads and public transport services;
- provide bus and rail services to give reasonable accessibility and a reasonable fare;
- increase the attractiveness of bus services by higher capital investment;
- improve the strategic highway network to attract longer distance and heavy goods traffic and reduce the impact of such traffic in residential and other unsuitable areas;
- bring roads in the older industrial areas up to modern standards;
- encourage adequate parking for shopping and business in the major commercial centres.

This was much less ambitious than the IPPS fifteen years earlier. Gone almost completely was the aim of rebuilding the strategic highway network as an urban motorway system. The only major proposals remaining in Liverpool were the inner ring road and the Garston by-pass. Investment in the local rail network was still being advocated, including electrification of the lines to Hooton, Hough Green, the Edge Hill Spur, and St Helens, but other schemes had been abandoned. There was some consideration of pedestrians and cyclists insomuch as special account was to be taken of their needs in the design of road proposals. The idea that parking policy might be used to control traffic generation appeared to have been abandoned with a statement that:

> The provision of adequate short-stay car parks in shopping and commercial centres will be encouraged…on-street parking in town and district centres will be permitted (Merseyside County Council, 1979, p117).

It was proposed that a further 125 hectares of land within Liverpool should be developed for public open space and sports pitches. The provision of further land for allotments was also a concern. Liverpool's inner areas were also identified as short of indoor sports facilities and swimming pools. Whilst the city had been exemplary in the building of municipal swimming baths in the late 19th century, many of these facilities were becoming obsolete or had already closed, leading to an acute shortage of modern provision.

The potential of the city as a tourist destination was beginning to be recognised with a modest policy to retain and enhance existing tourism assets and to protect the tourist potential of areas. In addition new attractions were to be encouraged, especially those on the Mersey waterfront, and including the interpretation and exhibition of maritime and industrial history, cultural and entertainment facilities, sports and recreation facilities, conference facilities and the provision of hotels.

Retail policies were to protect the status of Liverpool City Centre, guide retail investment towards existing centres, refuse consent for out-of-town shopping developments, and improve the retail environment. Liverpool's district centre policy remained in tact with Allerton, Belle Vale, Breck Road, Broadway, Edge Hill, Garston, Kirkdale, Old Swan, Toxteth and Walton Vale all being named as centres where large shopping developments would be permitted and which would be protected from competition from off-centre developments.

Implementation

Implementing the structure plan was always going to be much more complicated than the implementation of the 1965 IPPS. Local government had been reorganised and many functions had been split between county and district councils or hived off to new public utilities or regional bodies as shown in Table 5.2. Public spending was under increasing pressure and scrutiny. Local people were increasingly vociferous about planning matters and there were strong pressure groups emerging in pursuance of environmental and social concerns, while the property and development industries were finding a sympathetic political ear for their agenda.

The structure plan contained some policies that could be implemented by means of the County Council's own spending, for example on highway investment. It also contained some firm policies for the control of development that could be enforced as soon as the plan was approved, for example those policies that expressly permitted or forbade certain forms of development (e.g. out-of-town shopping centres would not generally be permitted). The proposal for a 'green belt' was translated into action through the Green Belt Local Plan (Merseyside CC, 1983).

Table 5.2   Implementing the Merseyside Structure Plan

Chapter	Policy	Main agency for implementation
Economy	Economic development	MCC, MERCADO, DC, MSC, MDHC, DOI
Housing	Treatment of obsolete housing	DC, HA, private owners
	Council housebuilding	DC
	Land for housebuilding	DC
Urban Environment	Reclaiming derelict land	MCC, DC
	Parks and Trees	DC
	Building conservation	DC
	Residential environment	DC
	Industrial environment	DC
	Waste disposal	MCC
Natural Resources and Open Land	Enhancing water quality	NWWA
	Canal improvements	BWB
	Minerals planning	MCC, NCB
	Rural economy	MCC, MAFF, NCC, DC, FC, CC
Transport	Passenger Transport	MPTE, BR
	Roads	MCC, DC, DOT
	Car parking	MCC, DC
	Cross-river facilities	MPTE
	Liverpool Airport	MCC
Recreation	Sports and recreation	DC
	Countryside recreation	DC, CC
Tourism	Tourism development	MCC, DC
Shopping	New development	DC, private developers
	Environmental works	DC

BR	British Rail	MCC	Merseyside County Council
BWB	British Waterways Board	MERCADO	Merseyside Economic
CC	Countryside Commission		Development Office
DC	District Councils	MSC	Manpower Services
			Commission
DOI	Department of Industry	MPTE	Merseyside Passenger
			Transport Executive
DOT	Department of Transport	NCB	National Coal Board
FC	Forestry Commission	NCC	Nature Conservancy Council
MAFF	Ministry of Agriculture	NWWA	North West Water Authority

Source: author

However, much of the strategy required public spending and the making of regulations that could only be implemented through negotiation with other authorities. For many structure plan policies to be translated into action links would have to be established with the plans and programmes of the district councils, for example, through the preparation of local plans and housing investment programmes. Many of the policies for economic development and social change were no more than aspirations that the County Council would 'encourage' others to implement, including the private sector, through advocacy and negotiation.

The structure plan was to be a guide for the development of Merseyside in the nineteen eighties. During that decade the tightly drawn Green Belt was fairly effectively held, although the pressure for development was admittedly sluggish for much of the period. The complementary urban regeneration policies also showed a degree of success. Sufficient housing land was found within existing urban areas to accommodate most of the conurbation's needs. The decade also saw a growing amount of private sector investment in the city centre, docklands redevelopment and the inner areas, although most of this was supported by public subsidies. More success was achieved with derelict land reclamation than had previously been the case, especially for hard (i.e. profit making) end uses. The programme of highway improvements was completed more or less as planned but the level of maintenance and repairs declined. Public transport usage and income was stabilised until the deregulation of bus services brought in by the Transport Act 1986.

One of the biggest inhibitions to successful implementation of urban regeneration was the inadequacy of the resources available. In real terms during the decade the overall level of public spending in many inner areas was actually falling. Furthermore, central government was gradually shifting public funding for regeneration from mainstream local government spending towards ad hoc agencies and special projects that they themselves could control. Thus local democratic power and control over the regeneration process was being reduced whilst the influence of central government, ad hoc agencies and private investors was increasing.

Conclusions

Almost as soon as the Town and Country Planning Act had hit the statute book the planning system began to be undermined by a series of alternative policy mechanisms that achieved similar ends and, in particular, by the reorganisation of local government in 1974. The effect of all of this was to fragment responsibility for different elements of urban policy and after 1974, to fragment the responsibility for planning itself.

However, during this same period there began to emerge two new concerns: urban deprivation and the need for urban regeneration, and the environmental agenda. Both, in very different ways, were to have a fundamental impact upon the planning system in the years to come.

Despite the turmoil and change of the period, two strategic plans did emerge to shape the planning framework for Liverpool. The first of these, the Strategic Plan for the North West, sought to concentrate regional development in the Mersey Belt between Liverpool and Manchester. This regional plan was also significant for its recognition of the emerging environmental agenda and the financial problems facing the big city authorities. Perhaps the biggest difficulty in implementing this plan was that this was regional planning without regional government, so that it was unclear how conflicting aims would be resolved and how the big decisions could actually be taken. The other plan was the Merseyside Structure Plan. This was a key document in establishing the twin policies of investing in the inner urban areas and restricting of peripheral growth as the basis for urban regeneration.

Table 5.3   The characteristics of planning and urban regeneration in Liverpool during the period of strategic planning

What was the relationship between planning and urban regeneration?	Emerging studies on urban deprivation were informing the planning process. Urban regeneration was becoming a central aim of planning policy.
What were the main achievements of planning and regeneration during the period?	Growing understanding of the nature of urban deprivation. New policies for urban regeneration and restricting urban sprawl. Emergence of a nascent environmental agenda in planning.
What was the relationship between economic, environmental and social aims?	Relatively balanced approach. The need for economic development, environmental protection and social inclusion were all represented in policy.
What was the extent of local democracy and participation?	'Partnership' between central and local government. Growing community participation in the planning process and in area improvement programmes and the urban deprivation studies and experiments.
To what extent were policies co-ordinated as part of long-term strategy of fragmented and short-term?	The Structure Plan nested fairly well within regional strategy. Both taking a long-term view. Some friction appearing between strategic (County) and local (District) policies.

Source: author

Table 5.3 gives an indication of the characteristics of planning and urban regeneration in this period. Compared with the modernisation period these plans were based upon a more rigorous analysis of the problems of the area and the dynamics of urban change. They broadened the planning agenda to give more consideration to economic development, social issues and the natural environment. The emergence of regeneration as the dominant policy aim gave the document a clear focus.

Bold as it was, one of the fundamental problems facing the regeneration strategy was the reliance on so many disparate agencies for its implementation. The scope of the plan, using land use control and development policies as its main tools, was too narrow to achieve urban regeneration, which required a wider, more comprehensive multi-agency strategy to succeed.

But the structure plan was not the only policy instrument driving urban regeneration. In 1978 the Labour Government had passed the Inner Areas Act and set up the Liverpool Inner Area Partnership. Two years later the new Thatcher Government had established the Merseyside Development Corporation with its brief to redevelop the former docklands and its ability to operate outside local democratic control with limited regard to local planning policies. Indeed the eighties became the era of market led regeneration with a strong anti-planning and anti local government hegemony which reached its zenith in 1986 with the publication of 'Action for Cities' the Thatcher Government's ultimate 'solution' to the crisis facing inner urban areas. The effect of these policies and agencies are discussed below in chapter six. The regional plan and structure plan represented only the upper tiers of the planning process. Within the strategy framework they provided it was still necessary for the City Council to prepare development plans. What plans emerged at the city scale is discussed below in chapter seven.

6 Property-led Regeneration

Margaret Thatcher came to power on 3rd May 1979, leading a government that combined a new right-wing conservative philosophy with elements of liberalism, populism and a degree of pragmatism. This new government had little time for the idea of regenerating urban areas in partnership with the inner city local authorities, most of whom were Labour controlled and held in rather low regard by many ministers. Whereas the late 1970s had been characterised by something approaching a consensus of view about urban policy between central and local government, the new government was much less trusting of the ability of local authorities to deliver regeneration and was more responsive than its predecessor to the concerns of private investors and property developers (Atkinson and Moon, 1994; Lawless, 1988; Robson, 1988).

The new government quickly reversed the Labour government's rate support grant changes and played down the role of the inner city partnerships in the regeneration process. According to Merseyside County Council the Liverpool Inner City Partnership effectively ceased to function in 1981 (with the last meeting held on 12th January 1981) when Michael Heseltine, Secretary of State for the Environment, took on his Merseyside role (see below). There was no longer any real co-ordination between 'partners' with the Department of the Environment subjecting urban programme submissions to 'detailed scrutiny, causing delay and preventing the development of effective rolling programmes' (Merseyside County Council, 1985, p7).

Instead the new government sought more direct forms of intervention that could be more easily maintained under its own control to achieve its own aims without the interference of local authorities. Under the provisions of the Local Government, Planning and Land Act 1980, the government established 'enterprise zones' to stimulate private investment; 'urban development corporations' to reclaim large tracts of derelict urban land; and 'registers of publicly owned vacant land' to encourage the disposal of surplus land from public authorities and utilities. Thus, with a strong emphasis on property-led regeneration, a system was put in place that increasingly by-passed local democratic control and transferred investment and development responsibilities to central government agencies and the private sector. Figure 6.1 indicates the main area-based regeneration policies and programmes in Liverpool during this period.

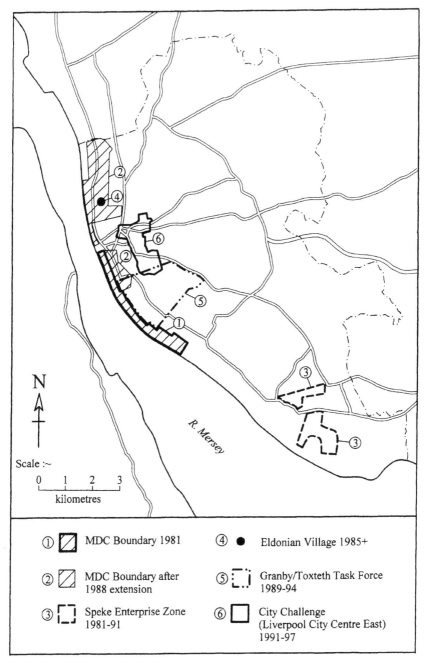

Figure 6.1   Area-based policies and programmes, 1981-91

Source: author

Speke Enterprise Zone

One of the first examples of the government's new approach to regeneration was the 'enterprise zones' (EZ). These were conceived as an experiment to explore the extent to which industrial and commercial regeneration might be promoted through designation of an area that, for ten years, would benefit from the streamlining of planning procedures and administrative controls and various tax advantages, notably exemptions against business rates and corporation tax. It was thought that in such zones entrepreneurial activity might be stimulated and that the areas might become seed beds for economic growth that would have multiplier effects within the local economy (Department of the Environment, 1988, p48).

Speke Enterprise Zone was designated in 1981 and included 133 hectares of land and buildings in two zones: the former British Leyland and Dunlop factories and an area of the 'old' Liverpool Airport which had become surplus due to the reorganisation of airport operations. The zone was subject to a simplified planning regime in which planning permission would normally be granted subject to conformity with a basic land-use zoning plan. Unlike most other EZ, Speke contained only a limited amount of dereliction and offered potential investors immediately developable land. Despite this advantage the area attracted only a modest amount of development. By the end of the decade the only major development was that carried out by a government agency: English Industrial Estates, whose programme of advance factory construction had actually begun prior to designation.

Between 1981 and 1989 the total private investment in the Speke EZ was less than £20 million, compared with around £40 million in Trafford Park, £100 million in Salford and over £170 million in Gateshead (Department of the Environment, 1994, p102). Nationally, it has been estimated that about 35,000 jobs were created between 1981 and 1986 by the EZ policy (PA Cambridge Economic Consultants, 1987) but few of these were located at Speke where, even at the end of the ten-year life of the EZ, much of the land remained vacant.

Merseyside Task Force

In the summer of 1981 riots broke out in a number of inner urban areas, including Toxteth in Liverpool, St Pauls in Bristol, Brixton in London and elsewhere. In the words of Brian Robson:

The major disturbances of 1981...focused popular and official attention on the plight of cities. The riots undoubtedly reflected genuine economic distress, the casual violence of an alienated population, and above all the often merited hatred felt alike by black and white youths towards the police as emblems of an alien authority (Robson, 1988, p37).

The riots in Toxteth were particularly sustained and violent. This led to a high profile response from the government, which included Michael Heseltine (Secretary of State for the Environment) being, for a short period, styled the 'Minister for Merseyside' and setting up the Merseyside Task Force (MTF) together with a series of national policy initiatives designed to get more private sector investment into the inner cities.

Figure 6.2   Old and new housing at the Rialto
The Rialto building was an important landmark in the inner city and at the epi-centre of the 1981 Toxteth riots. Burned down at the time, it has been replaced by a new building containing shops and offices for a social housing organisation

Source: author

The MTF was established in October 1981. Although intended originally to be in existence for only one year it remained until 1993 when it was recast as the Government Office for Merseyside. The MTF included civil servants from the Department of the Environment, Department of Industry, the Manpower Services Commission, and the Department of

Transport. Its role was to co-ordinate government policies and spending programmes in the area and to generate new initiatives. It was particularly concerned with the reduction of unemployment (which stood at around 20 per cent in the city at that time) and improvements to the economic and social life of the conurbation. There was also mention of the need to promote good community relations and reduce disadvantage among ethnic minority groups (Central Office of Information, 1981).

It was only after the Toxteth riots that local planning and urban regeneration policies began seriously to consider the need to combat racism and to take account of the needs of ethnic minorities in policy making. The Merseyside Structure Plan, approved only months before the riots, made no mention of these issues. Nevertheless, it has been suggested that even though the MTF's mission emphasised racial issues, they quickly became marginalized.

A few 'projects' to benefit the black community were ultimately introduced. But they were secondary to the broader Merseyside-wide initiatives which were almost entirely irrelevant to the black population (Ben-Tovim, 1988, p144).

The County Council too was critical of the MTF:

Merseyside Task Force has itself failed to realise its original far-reaching brief which was to do with the overall deployment and effectiveness of public resources in Merseyside and the need to modify policies or switch resources between government departments. This focus was soon lost in a plethora of local initiatives which Merseyside Task Force have pursued on a project by project basis. Whilst these special initiatives are important in their own right, they have not been a substitute for the better co-ordination of policies and programmes...or the search for more fundamental solutions to Merseyside's problems (Merseyside County Council, 1985, p7).

Despite these criticisms, as the conduit through which much government funding was to be channelled, the MTF played an increasingly important role in the regeneration of the city, promoting training projects, housing improvements, tourism and leisure projects, encouraging small business development and other initiatives (House of Commons Environment Committee, 1983, p6).

One example of its influence was in the development of the Wavertree Technology Park. The technology park had originally been a joint initiative of the former Merseyside County Council, British Rail, the English Industrial Estates Corporation and Plessey (the electrical engineering firm) to regenerate derelict former railway sidings in the

Wavertree area as a high-technology industrial campus. Supported by the MTF work started in 1984, development of the original site of 26 hectares

Figure 6.3   **Wavertree Technology Park**
A new building completed for the Sony Corporation. The landscaped technology park is in marked contrast to the industrial landscape that used to characterise the area

Source: author

was so rapid that by 1988 a second phase was added comprising a further 9 hectares of adjoining land. Investments were attracted from Barclaycard, (creating 600 new jobs), Plessey Crypto (communications), Powell and Schofield (biochemicals) and a number of other smaller firms. The park was a great success as a location for high-technology industries and others seeking to benefit from that image, with expansion continuing throughout the nineteen nineties. To improve access a new Wavertree Technology Park railway station opened in 2001.

Another important national policy initiative to emerge from Michael Heseltine's ministry in the wake of this 'official focus on the plight of cities' was Urban Development Grant (UDG). Borrowing an idea from the United States, UDG was launched in 1982 with the aim of promoting economic and physical regeneration in run down urban areas by encouraging private development that would not otherwise take place. Property developers had, quite reasonably, argued that they could not be expected to invest in areas of low demand where the costs of development exceeded the return they could obtain for the completed building. The purpose of UDG was to bridge the gap between the cost of a development and its value on completion. UDG and its successor, City Grant, played an important role in subsidising developments in Liverpool over the next decade.

Political Conflict in the 1980s

The mid-1980s were a period of local political turmoil in Liverpool. It was a time when the Labour controlled City Council was in deep conflict with Margaret Thatcher's government. The effects did long-term damage the council's ability to provide adequate local services and seriously undermined any attempt to attract inward investment during the late 1980s and early 1990s. The events have been well documented by Parkinson (1985) and (1988). Liverpool's economic situation had been steadily worsening through the 1970s as government initiatives failed to tackle the underlying problems. Between 1974 and 1983 no political party had been able to maintain an overall majority on Liverpool City Council, although the Liberal party managed to hold the balance of power through much of the period and were influential in limiting both spending and rate rises. It was a period when the council found it difficult to set budgets and difficult to agree policies and programmes.

After the Conservative victory in the general election of 1979 the Liverpool Labour party began to move from its traditional right of centre position to adopt a hard left 'militant' ideology. In 1983 the local Labour party gained overall control of the City Council for the first time in over ten years and the scene was set for a major political confrontation. A decade of indecision had left the City Council with serious financial problems that were made worse by the Conservative government's new grant regime. The Council felt that the city was being unfairly treated and in 1984 confronted the government with a threat to bankrupt the council if they would not give Liverpool more money. The threat was abandoned only after some creative accounting by the City Council and some minor concessions from the government. During 1985 the tactic was tried again but the government made no further concessions although further creative accounting again saved the council from the immediate threat of bankruptcy. However, because of the financial irregularities involved, in June 1985 some 48 local Labour councillors were surcharged and disqualified from office by the District Auditor. The local Labour party found itself increasingly isolated from the party nationally. The National

Executive Committee of the Labour Party suspended the Liverpool City party and in 1986 expelled several leading members of the local party (Parkinson, 1988, p110 et seq.).

Only after the local elections in 1987 was there a return to something like 'normal' politics when, despite the disqualification of the 47 councillors, the Labour party was returned to office in the city with an even bigger majority than before. However, most of the new councillors were closer to the political centre and opposed to the tactics of the previous 'militant' administration and the politics of confrontation were abandoned (Parkinson, 1990, p252).

Through the early 1980s Merseyside County Council had worked hard on economic development campaigns to improve the image of the area as a location for inward. Merseyside Development Corporation (see below) had also made a positive contribution through its support of the International Garden Festival, the Tall Ships Race (1984) and the restoration of the Albert Dock, which became an icon for the regeneration of the city. But by the end of this period the city had become a by-ward for political turmoil and militancy: not an attractive destination for footloose capitalist investment. Whatever gains the 'militants' thought they might have made were almost certainly dwarfed by the lost investment and poverty of council services over the subsequent decade.

Merseyside Development Corporation

The Merseyside Development Corporation (MDC) was established in 1980 with a remit to redevelop parts of Merseyside including the South Docks. But this was not the first attempt to tackle the problem of Liverpool's derelict docklands for nearly a decade earlier the City Council had published 'Liverpool South Docks: Principles of Redevelopment' (Liverpool City Planning Department, 1972).

The objective of that report was to set out the planning principles that should guide redevelopment in order to maximise economic and environmental benefits whilst avoiding 'undesirable interim conditions'. It was proposed that redevelopment should include 'employment-creating' development, integrate the area with its surroundings, open up the riverfront to public access, ameliorate the local shortage of open space; provide for water-based recreation and promote 'civic and metropolitan uses'. Three main land uses were proposed: industry and warehousing, to provide employment opportunities in the southern part of the inner areas; housing, to exploit the opportunity 'to create new riverside residential areas for a variety of income groups'; and up to 18 hectares of open space to meet the deficiencies of the surrounding residential areas. A fourth land use, offices, was only to be encouraged after existing sites in the city centre had been developed. An interesting aspect of the proposals was the way in which they formed part of an integrated process of planning the city: contributing towards solving the problems of dereliction whilst not competing for the same resources and investment as other areas.

By 1975 a pressure group (the Docklands Action Group) had been formed and produced its own ideas for the future. This group sought not to produce a masterplan which 'would produce only a lifeless environment' but a set of principles. These were essentially socialistic and based on awakening local interest in the potential of the docklands, encouraging public participation and using co-operatives and partnerships to redevelop the area for the benefit of local people. Their proposals included industry, warehousing, residential developments to meet local needs, and new types of employment that utilised the features and character of the area, including:

- using a dockside shed as a transport museum;
- converting the Queens graving dock into a sports stadium;
- a fish farm in Coburg Dock;
- recreational moorings,
- boat building and repairs in Brunswick dock;
- using Harrington Dock for water sports;
- using Herculaneum Dock as a marina for use by local schools.

It was also anticipated that the site had characteristics, including a high tidal range and high wind speeds, that could be exploited to generate renewable energy supplies (Docklands Action Group, 1975). Today, many of the group's proposals would be regarded as mainstream to theories of sustainable development. In 1975 these were novel ideas indeed.

By the late 1970s Merseyside County Council had its own ideas about docklands regeneration and in 1979 published its 'South Docks Prospectus'. As a first step towards regeneration the County Council had decided to buy the South Docks (Merseyside County Council, 1979, p1). The prospectus gave an indication of how the docklands could be redeveloped. Canning, Salthouse and Albert Docks would house a maritime museum to be developed by the County Council together with a 'trade, industry and export centre' (exhibition halls, shopping mall, offices, studios and flats). Kings and Wapping Docks would accommodate a commercial development including a 12,000 m^2 hypermarket and refurbishment of the Wapping Dock Warehouse for offices. At Queens and Brunswick Docks provision for water sports with commercial or residential development was envisaged. The area around Toxteth dock was to be used for industrial development and finally an enclave of housing was to be built on a small inland site at the southern end of the area.

Apart from the maritime museum, which was by then a firm County Council proposal, it could be said that this was not a well worked out redevelopment plan but merely a collection of possibilities. There was little or no justification for the proposals and little consideration of costs, funding or the implementation process. However, the County Council was never given the chance to acquire the land nor to carry out its redevelopment plans as the incoming Conservative government took the view that a single-minded development agency would be a more appropriate vehicle for large scale reclamation and regeneration schemes such as this. It was felt that such an agency would have a better understanding of property development processes, take decisions more quickly and create a better and more stable investment climate than could be achieved by a local authority.

The Local Government, Planning and Land Act 1980 provided the Secretary of State for the Environment with the powers to designate areas and establish Urban Development Corporations (UDC) for the purpose of area regeneration through the reclamation of derelict property, encouraging industrial and commercial development and ensuring the provision of social facilities and housing. These central government agencies were to comprise a Board accountable to the Secretary of State with a chief executive and staff responsible to the Board. The action-orientated structure and style was more like that of a development company than a local authority. They had powers to acquire, manage and dispose of land, to carry out reclamation works and to provide infrastructure for development. They also had powers of development control within their designated area.

The first two UDCs were established in 1981 in the London docklands and on Merseyside. In Merseyside the designated area comprised 350 hectares including the former Liverpool South Docks, parts of the northern docks in Sefton and land on the Wirral side of the Mersey. Although sceptical and concerned about the lack of local accountability, the City Council did not formally object to its establishment (Hayes, 1987, para. 15).

The first plan published by the MDC was the 'Initial Development Strategy' (IDS) (Merseyside Development Corporation, 1981). A slim document of only 25 pages, its purpose was to set out a reclamation strategy and identify the land uses to remain, or be attracted to the area. It was to be a flexible framework for public and private sector investment, a guide to the control of development and a programme for land acquisition and reclamation. It claimed, curiously, to be neither a non-statutory land-use plan nor a master plan. Their analysis was that:

> The whole environment has become severely degraded since the closure of the docks in 1972 when they became tidal. Contaminated silt is now up to 30 feet deep and is exposed except for 2 or 3 hours on each tide. The large historic buildings in the north remain almost unused and are falling rapidly into disrepair. In the Brunswick/Coburg Docks small under-capitalised business activities, though providing valuable employment, fail to maintain premises and suffer from lack of services to deal with waste disposal and other needs (Merseyside Development Corporation, 1981, p17).

MDC therefore saw its objectives as:

- securing the regeneration of the area;
- bringing land and buildings back into effective use;
- encouraging commercial and industrial development;
- creating an attractive environment that would encourage people to live and work in the area.

It was felt that the most important single action that could be taken to improve the physical environment was to restore water permanently to the docks. Construction of the inner ring road would improve access to the north end of the area, whilst the rebuilding of Sefton Street (the dock road) from Park Street southwards towards Herculaneum Dock would improve access and 'provide a landscaped frontage to new industrial development'.

On the north side of Canning Dock, a site was identified for a large office development that could be 'linked by pedestrian bridge across the Strand to adjoining office and commercial areas'. No such office development has yet taken place and no pedestrian bridge has been constructed, despite numerous studies and proposals. Pedestrian access into the area from the city centre remains problematic to this day. The strategy recognised the architectural and historic importance of the Albert Dock warehouses but at that time was unable to put forward any firm comprehensive proposals for their re-use other than the maritime museum. Whilst commercial development might be attracted together with recreational and residential accommodation, it was thought 'unlikely that substantial shopping floorspace could be built without detriment to the Liverpool city centre'. To this degree the strategy showed a sensitivity to

local economic and planning issues that had been absent from the previous
County Council proposals.

The area from the Albert Dock southwards to the Queens Dock was to
be redeveloped as a zone of mixed commercial, recreational and residential
activities. The whole of the remaining docks that lay further from the city
centre (Coburg, Brunswick, Toxteth, Harrington and Herculaneum) were
earmarked for industrial uses. At this stage it was anticipated that water
would be retained in most of the docks except Harrington and Herculaneum
although subsequently decisions were taken to fill in more of these
southern docks so as to provide a greater amount of useful land for industry
and parking. A pedestrian walk was to be constructed along the whole
length of the riverside, turning inland only around the Coburg/Brunswick
Dock granaries, which were considered too expensive to relocate.

South of Herculaneum Dock the Riverside area was virtually
inaccessible from the landward side. The area comprised a disused and
part-cleared tank farm, a local authority tip, neglected woodlands and
playing fields. Yet the area adjoined sought-after residential areas and
commanded extensive views over the river. A key proposal was to improve
access by extending the reconstructed Sefton Street southwards through the
site to rejoin the existing primary network in Aigburth. Not only would
this proposal open up the land for development but would also provide a
new through route to the city centre from the southern suburbs. This area
was thought to be suitable for development as a 'science park', up to 800
dwellings and public open space. The strategy contained no mention of the
'garden festival' that was to be proposed very soon after its publication.

One of the first priorities for the MDC was the restoration of the water
regime to the South Docks and the creation of a major parkland area at
Riverside. At the same time local environmental projects were to be
undertaken to improve the appearance of the area whilst derelict buildings
were to be removed or refurbished.

Early progress was impressive and in an evaluation report the DOE
commented:

> In the MDC area £140 million of public investment has helped to reclaim 97
> hectares for residential and commercial development and 48 hectares for
> recreation and open space; and to refurbish 135,000 square metres for
> housing and commercial uses, including the historic Albert Dock restoration.
> The MDC has also created 1,160 jobs since 1981 and 94 per cent of its
> contracts have been let to firms in the Merseyside area (Department of the
> Environment, 1988, pp52-53).

Writing in 1987, Michael Hayes, then Liverpool's City Planning Officer, provided one of the first detailed evaluations of the MDC and its achievements. In refining the IDS, the MDC had appointed consultants who recommended that the full potential of the waterfront could be best achieved through a tourism and leisure development strategy. The MDC accepted this view, partly because by 1984 the International Garden Festival, the Tall Ships Race and the redeveloped Albert Dock were already demonstrating the drawing power of the area. A series of major leisure projects were proposed including an aquarium, water-park, ice-skating arena, hotel and exhibition centre (Hayes, 1987, para. 25). Various problems including limited local demand, political uncertainty and site-specific design difficulties resulted in none of these ambitious projects being implemented, with the exception of a small number of hotels. This is not to say that the consultant's analysis was necessarily wrong. A large aquarium was built on the other side of the Mersey at Ellesmere Port in the mid-nineties. An ice-skating arena was built in Manchester (MEN Arena) and multiplex cinemas have been developed elsewhere in the city, but the continuing lack of a major exhibition and conference centre in Liverpool remains an issue.

Lack of demand for industrial and office space led to other changes to the IDS, notably reallocating the former Dingle Tank Farm site for housing and the proposal to redevelop the Herculaneum dock for retail warehouses rather than as a business park (Hayes, 1987, p25). This point is reinforced by Meegan (1993) who suggested that the goal of attracting inward investment was always going to be difficult because of the fragility of the local economy, local political instability and the poor image of the dockland environment at that time. This led to a rethinking of the initial strategy with more emphasis placed on housing and tertiary employment rather than industrial development as was originally envisaged (Meegan, 1993, p61).

By the mid-1980s the restoration of the water regime in the dock system was nearing completion. In other docks infilling had been completed. In the area of the former Toxteth and Harrington Docks a combination of new and refurbished buildings was providing new industrial units, warehousing and 'managed workspaces' for new firms. A housing association was developing 100 dwellings for rent in the Dingle Mount area. Barratt (the housebuilders) were converting the Wapping Dock warehouse into private flats for sale and the developers Arrowcroft were providing flats as part of the redevelopment of the Albert Dock (Hayes, 1987, paras 29 and 30).

Figure 6.4   The Albert Dock
The dock basin with the metropolitan cathedral in the
background. It was the restoration of the Albert Dock that kick-
started dockland regeneration in the city

Source: author

At this time, UDCs were frequently criticised for their lack of local
accountability, lack of co-ordination with other local agencies and in some
circumstances, the absence of satisfactory forward planning. Two
controversial development control decisions illustrate the tension between
the MDC and the local authorities. In 1982 the 'Pavilion' proposal to
develop a major shopping and recreation centre at the Kings Dock (which
had earlier been supported by the Merseyside County Council in their
South Docks Prospectus), was opposed by the City Council on the grounds
that it would have a detrimental impact on the city centre. Despite their
earlier statement in the IDS, the MDC rejected this view and granted
planning permission. Ironically the developers subsequently abandoned the
scheme. In 1987 the MDC again approved a large retail warehouse scheme
at Herculaneum Dock despite strong objections from the City Council who
had urged the Secretary of State to call-in the application. He chose not to
do so and supported the MDC in its decision. In the end, largely for
reasons of financial viability, the development did not go ahead (Hayes,
1987, para. 48).

The experience of the Liverpool International Garden Festival illustrated the consequences of unsatisfactory forward planning. The notion of a 'garden festival' had been based upon a German idea, where such events were held from time to time in different cities both as a celebration of gardening and landscape arts and as a mechanism for reclaiming derelict land for subsequent re-use. The proposal for an international garden festival in Liverpool was supported by Michael Heseltine but only received government approval towards the end of 1981 and had not formed part of the original IDS. Nevertheless the MDC adapted its plans and programmes and worked with impressive speed to host the Liverpool International Garden Festival during the summer of 1984. Some 93 hectares of land were reclaimed, of which 49 hectares were used for the festival itself. The cost of land acquisition and reclamation was £14.1 million and the cost of developing the site as a garden festival, was £18.9 million (both at 1990 prices). The event itself was a huge success, receiving critical acclaim and attracting in excess of 3.4 million visitors (Department of the Environment, 1994, p125).

However, the organisers of the International Garden Festival had pressed ahead in the knowledge that no proper arrangements had been made for the continued funding and operation of the gardens after the end of the festival that year. This led to major problems. In 1983 the City Council had anticipated that the Festival Hall (the central hall of the garden festival and designed by Arup Associates) would be converted into a major new leisure centre (including sports facilities and leisure pool) and fill a marked gap in the city's recreational facilities (Liverpool City Planning Department, 1983, para. 8.25). Instead the site was let to a private company to run as a leisure attraction. After two years the company went into liquidation (Hayes, 1987, p29). After one or two further attempts at temporary uses the site was closed and mothballed. Even in 2002, apart from the selling off of further parts of the festival site for housing, little had changed. What should have been an asset to the local community has become instead a monument to the consequences of a failure to think through and plan the long term use of the site.

'Action for Cities'

The 1987 general election brought an easy victory to the Conservatives. Only in the inner urban areas did voters prefer the Labour alternative. According to Robson:

In the early hours of the morning after the General Election in 1987 and ebullient from her victory, the Prime Minister stood on the stairs of Conservative Central Office to address her party workers. Her first words were about the inner cities. 'So no one must slack....on Monday we have a big job to do in some of those inner cities'. Whether this was meant as a presage of a new policy or merely a political message is open to interpretation. What is certain is that, for a moment at least, the media were to focus on the inner cities as the matter of the moment and the government, in its turn, gave sustained attention to (the) issue (Robson B, 1988, p vii).

Kenneth Clarke was appointed as Minister for the Inner Cities and shortly afterwards the government produced a new strategy document: 'Action for Cities' (Cabinet Office, 1988). This document represented the lowest point in the Conservative government's scepticism towards the competence and value of local authorities in the regeneration process and perhaps the highest point in the government's view that inner city policy was predominantly about property development and the maintenance of investor confidence. Within the 32-page document local authorities were mentioned only eight times, mainly in a critical or negative manner. Of the policies mentioned virtually all were put forward as central government initiatives whilst successes were claimed as resulting from central government and its agents or its private sector partners. Partnership had come to mean not a relationship between central and local government but a relationship between the state, usually in the form of a central government agency, and a private developer.

It was in this political climate that in November 1988 the designated area of the MDC was almost trebled in size to include, amongst other areas, part of Liverpool's northern docks together with their immediate hinterland in the Vauxhall and Kirkdale districts, and the south-eastern quarter of the city centre lying directly behind the south docks. One of the immediate consequences of this extension was that the MDC, for the first time, had to deal with areas of the city that were already densely developed and home to a significant indigenous population and an active economy. Such a situation required a much more participative approach to planning and development than had previously been thought necessary. This opened up a second phase in the life of the MDC.

In 1990 the MDC produced a new development strategy together with detailed local area proposals. The purpose of the new development strategy was spelled out as 'comprehensive regeneration', in order to make these areas part of the city again and help Liverpool recapture its status as an international city (Merseyside Development Corporation, 1990a, p4). The aims of the strategy were to:

- encourage enterprise, new businesses and help existing businesses to grow;
- improve people's job prospects, their motivation and skills;
- create a better environment for residents and businesses;
- make inner areas more attractive places in which to live and work;
- market Merseyside to potential investors, businessmen and tourists.

Figure 6.5   **Waterloo Dock**
The former warehouse has been converted into flats. New housing has also been built on surplus land between the dock and the River Mersey. Situated about one kilometre north of the Pier Head and poorly served with local amenities the regeneration of this area took a long time to achieve

Source: author

These were much more ambitious and grandiose aims than those of the IDS back in 1981 where the major concern had been limited to bringing derelict areas back into beneficial use. Now the emphasis was firmly on economic development and support for private investment. The MDC was being expected to play on a bigger stage and influence the future of the whole city. On this basis the MDC's planning proposals extended beyond their designated area to include comments on land use within the wider city centre and the strategic transport network throughout the city (Merseyside Development Corporation, 1990a, p16). Whether this was a legitimate function of the MDC is debatable but it is one illustration of how the MDC could attempt to influence the local planning process to suit its own property and economic development objectives.

Much of the extended designated area behind the North Docks was industrial in character.

> This is one of the city's main industrial areas with about 250 businesses. About 25% of the area, however, is vacant or grossly underused and the environment is sullied by a preponderance of low grade and waste-related uses, a number of which are in conspicuous location (Merseyside Development Corporation 1990a, p18).

In an approach reminiscent of land use zoning from an earlier era, the strategy of the MDC was to remove and concentrate waste-related activity on a few chosen locations and to improve the appearance of the area. Economic development was to commence with commercial development at the Princes Dock, just beyond the Pier Head, and to the north, industrial development would be attracted through a programme of environmental improvements along the main A565, marketed under the epithet 'Atlantic Avenue'.

Figure 6.6   **Princes Dock**
              Prestigious new offices, a conference centre and hotel have been
              built around the former Belfast ferry berth

Source: author

The Eldonians

Only in Vauxhall was the policy different. Here the plan was to build on the success of the Eldonian community and extend residential development along the Leeds-Liverpool Canal, providing up to 650 new dwellings. In the late seventies a group of residents from the Portland Gardens area of Vauxhall had formed a co-operative to develop their own housing as a replacement for cleared dwellings. Emboldened by this success, residents from the neighbouring Eldon Street sought to redevelop the site of the recently closed Tate and Lyle sugar refinery to provide new homes for local people. By 1982 was in the ownership of English Estates and by 1983 the Eldonian Cooperative had an option to purchase about 5 hectares of land on the site. It took until 1985 to obtain planning permission, not least because of 'militant' City Council's ideological objections to the idea of cooperative housing. Layout and design proposals came from the cooperative members themselves, working in collaboration with local architects. It took more than £2 million (mainly from Derelict Land Grant and the Merseyside Task Force) to get the derelict and contaminated land into a developable state.

Figure 6.7   **The Eldonian Village**
This elderly persons accommodation was provided as part of the Eldonian housing co-operative development
Source: author

The first 145 dwellings were completed in 1990. The extension of the designated area of the MDC provided an opportunity to extend the development from this first site towards the Leeds-Liverpool canal on the 'land to the north'. Here an additional 150 dwellings were built by Liverpool Housing Trust, on behalf of the cooperative. This was shortly followed by yet another development 'Athol Village', this time sponsored by the Vauxhall Neighbourhood Council. According to the MDC:

> Residents of the former Athol Street and surrounding areas were keen to get involved. They got together and linked with Merseyside Improved Houses which secured funding for the project from the Housing Corporation and acted as agents for the 150 new homes (Merseyside Development Corporation, 1997, p4).

By this time the private sector felt sufficiently secure to invest in new housing for owner occupation alongside the canal. By the turn of the century there was an enclave of more than 500 modern detached and semi-detached houses in an area that 30 years earlier had been a major industrial zone. Not only were houses built. A true community has emerged with sports and social facilities, retailing and even a garden centre. The whole development demonstrated not only the possibility of successfully redeveloping even the most depressed areas of industrial dereliction but also the ability of local communities to plan and design change for themselves. Here was proof that properly facilitated bottom-up approaches to regeneration could and did work.

Proposals for the so-called 'Liverpool Waterfront' area were much more commercial in nature. It was claimed that the area had the capacity to accommodate an expansion of the central shopping core and the main office quarter. It was no doubt correct to argue that the area had this 'capacity' and it was also true that the area would be attractive to developers because of the available of large, relatively cheap sites, good road access and car parking. However, no reasoned justification was presented to explain how this competition would benefit the city centre at a time when it already contained a large number of vacant sites and buildings with their own potential for redevelopment as retail or office units. What was happening was that the planning of the MDC's designated area was focused upon that organisation's objectives. It was detached from and not integrated with the planning of the rest of the city or the wider needs of the community.

The MDC argued that the combined effect of their initiatives would be greater than the sum of the parts and would lead to a more vibrant city centre with stronger links to the waterfront (Merseyside Development

Corporation, 1990b, p7). The MDC did extend the effective area of the city centre to include Princes Dock, Albert Dock and beyond but whether this added to rather than subtracted from the economy of the city centre is debatable. If the MDC had attracted to its area commercial developments that would otherwise have located in the city centre then the effect would almost certainly have been negative, tending to reduce property values and increase vacancy in the rest of the city centre. Only if the MDC had attracted developments that would not otherwise have come to the city could it have been argued that the effect would have been positive. The reality was that despite its ambitions, the MDC attracted so little commercial investment that it had little impact on the city centre one way or the other. By the millennium some industrial development had taken place in the Great Howard Street and Leeds Street areas. Residential development, including conversion of the former warehouses, was slowly being implemented in the Waterloo Dock area. Princes Dock had been partially redeveloped with a luxury hotel, conference centre and office buildings.

Figure 6.8 **The new Liverpool Marina and associated housing**
The marina, based on the former Coburg and Brunswick docks has proved an attractive location for new private housing investment

Source: author

The Pier Head had been remodelled and landscaped as a major civic space although access to this area from the city centre remained a problem. The Strand remained a six-lane highway that formed part of the city's inner ring road and represented a major barrier separating the Pier Head, Princes Dock and Albert Dock from the rest of the city centre. Despite the rhetoric of 'ensuring that the redeveloped docklands again became part of the fabric of the city' neither the MDC nor the City Council had implemented any solution in two decades of regeneration. It appeared that neither could resources be found to build a bridge nor could the alternative, a reduction in road width, be deemed acceptable.

Most of the refurbishment of the Albert Dock was completed during the eighties but the future of the Kings Dock site remained unresolved a decade later and by 2002 it was the only major site south of the Pier Head not to have been redeveloped. Offices and budget hotels were successfully realised around the Queens Dock and the redevelopment of the Coburg Dock and Brunswick Dock as a marina and residential area was impressive. Brunswick railway station opened in the mid-1990s. Virtually all the Sefton Street frontages had been redeveloped for low intensity uses such as motor vehicle showrooms and drive-in restaurants, a new railway station had opened at Brunswick dock and the nearby business park prospered.

Conclusions

As shown below in Table 6.1, the approach of the Conservative government to urban regeneration differed from that of its predecessor in a number of ways. It put less trust in the ability of local government to implement policy and put more emphasis on providing a robust framework that would encourage private investment. Although the previous Labour government had also recognised the importance of stimulating local economies through private investment, the mechanisms were different.

The Conservative government also gave more priority to the re-use of vacant and derelict urban land and the use of private-led property development as the mechanism to achieve this goal. This meant bypassing local democractic structures in order to ensure that the decisions taken not only accorded with central government policy but were also taken with the speed expected by private developers.

Much of this would have been acceptable had it been additional too, rather than a substitution for the efforts of the local authorities. The sharp cuts in rate support grant experienced in the early nineteen eighties forced inner city local authorities to cut back on many of their regeneration initiatives and to reduce spending on mainstream programmes, such as education, housing and social services, to the detriment of inner city residents.

Table 6.1   The nature of planning and urban regeneration in Liverpool during the period of property-led regeneration

What was the relationship between planning and urban regeneration?	The planning system began to be seen as part of the 'problem' inhibiting economic development. Urban regeneration became increasingly detached from the planning process (MDC, EZ).
What were the main achievements of planning and regeneration during the period?	Some re-use of vacant and derelict land and buildings. Successful area improvement and housing renovation programmes in the inner areas. Increasingly effective controls over urban sprawl.
What was the relationship between economic, environmental and social aims?	Economic development dominated political debate and action. Investment in land and buildings particularly encouraged. Environmental enhancements were justified on the basis of supporting economic development.
What was the extent of local democracy and participation?	Severe fracture in relationships between central and local government. Central government increasingly used direct action, non-democratic agencies and the private sector to implement urban regeneration, and encouraged community participation as a mechanism for bypassing local government.
To what extent were policies co-ordinated as part of long-term strategy of fragmented and short-term?	Many area-based or place-based urban regeneration programmes. Proposals tended to be shaped by the characteristics of the area or place with little strategic co-ordination.

Source: author

This approach to policy brought conflict. In 1981 disaffected youth rioted against the police and other symbols of authority in some inner city areas. Some local authorities vigorously objected to the government's urban strategy and cost-cutting approach to public services. In Liverpool the outcome was the surcharging of councillors. In other areas reforms were brought in by central government (e.g. Streamlining the Cities) to prevent subversion of their agenda.

Some government agencies, notably the LDDC, were faced with vociferous local opposition to their actions. Some experimental policies were also criticised. Speke Enterprise Zone exemplified the failure of that policy to deliver significant regeneration in areas were the local economy was weak. Nevertheless, there were also successes. Locally the MTF brought a useful element of co-ordination to government decision-making in the area. Wavertree Technology Park proved to be a successful investment with significant local economic benefits. The co-operative housing movement, and the 'Vauxhall villages' in particular, demonstrated the validity of bottom-up approaches to housing development. Overall the MDC achieved much of what it set out to do in bringing the formerly derelict docklands back into beneficial use. The Albert Dock became a major tourist attraction and the successful introduction of private housing to the areas around the new Liverpool marina was a notable achievement.

However, these approaches to urban regeneration, with their emphasis on ad-hoc initiatives and opportunism, tended to have little regard to wider plans for the development of the city. The evolution of those plans during the 1980s and 1990s is considered in the next chapter.

7   A Return to Comprehensive Planning?

Current Planning Policies and Development Programmes: Liverpool 1983

The delays in structure plan preparation had frustrated the City Council's own local planning ambitions. Indeed the city produced no statutory plans at all during the 1970s. Nevertheless, the city planners had a tradition of publishing substantial and comprehensive *informal* plans as a mechanism for guiding planning decisions. The Interim Planning Policy Statement of 1965 was one such plan. In 1983 another informal plan emerged. According to the Council this did not constitute a 'plan' in the conventional sense. Nevertheless, the document sought to:

- develop the policies of the Structure Plan
- provide a framework for local planning and development control
- assist in the co-ordination of development
- highlight parts of the Council's urban regeneration strategy
- act as a basis for public information and consultation
- provide a starting point for policy evaluation

By any reasonable definition therefore this was a 'plan', despite the protestations of the Council.

Much had happened over the previous decade to warrant the production of a new plan. The recession after 1973 and the reorganisation of local government in 1974 changed the political landscape. Urban regeneration had become the dominant concern of urban policy. There were growing restrictions on local authority expenditure. The planning system was now more fragmented with the County Council and other new agencies, notably the MDC, obtaining planning and development powers.

Since 1971 unemployment in the city had more than trebled from under 19,976 to 62,595 (the unemployment rate rising from 5.6 per cent to 20.2 per cent) (Liverpool City Planning Department, 1983, para, 4.2). Over the same period investment in commerce and service industries had failed to compensate for the continuing decline in manufacturing industry. The Council claimed that its economic regeneration strategy was therefore based on recognition of the increasing importance of the service sector, especially 'white collar' services to the future of the city. However, the majority of planning policies were still aimed at the supply of industrial land and improvements to existing industrial areas. The plan was downbeat about its ability to stimulate office employment:

> Future prospects for office development are largely dependent on the private sector. The City Council's ability to intervene and encourage the supply is limited (Liverpool City Planning Department, 1983, para.4.36).

The whole section on economic development was narrow in its conception and offered little by way of policies to support employment sectors such as retailing, leisure, tourism, health and education. Given that the Council had recognised the importance of the service sector to the future employment prospects of the city this was a significant omission.

Unsatisfactory quality rather than quantity had by now become the dominant housing problem facing the city. Since 1976 there had been a comprehensive strategy for renewing the city's older (pre-1919) private housing stock and good progress was being made. Buoyed by the provisions of the Housing Act 1974, the City Council had by now declared 70 Housing Action Areas (24,360 dwellings) and 31 General Improvement Areas (8,700 dwellings). Throughout the period this unsung but effective policy led to the renovation of an average of around 2,000 dwellings a year and made an important contribution to improving the housing conditions of residents in the inner areas.

But on the other hand the problem of unsatisfactory and 'hard-to-let' council housing was a new phenomenon that had only recently begun to emerge and as yet policy was experimental and piecemeal. The problem had a number of dimensions: some of the earliest inter-war council dwellings were becoming obsolete in terms of form (walk-up flats without lifts) and amenities (inadequate kitchens, bathrooms, heating systems). Some newer housing from the post-war period had been poorly designed and/or constructed. Much of the stock had been poorly maintained and some areas were becoming stigmatised through crime or social deprivation. Above all a growing surplus of dwellings over households permitted tenants greater choice and led to the least popular dwellings becoming increasingly difficult to let. Furthermore, an ageing population was in need of more 'special needs' and 'elderly persons' accommodation.

In March 1983 the Council had approved a new strategy for renewing the council housing stock. The five-year programme was intended to deal with 16,564 council dwellings in 34 priority areas through a combination of demolitions, conversions, improvements and repairs (Liverpool City

Council, 1983). The new Conservative government added another dimension to the housing issue by strongly encouraging home ownership as the preferred tenure. In pursuit of this local authorities were discouraged from building new council housing and were obliged to sell council dwellings to sitting tenants who wished to buy.

Through a combination of peripheral sites, infilling and brownfield sites (derelict land) it was estimated that the city had some 455 hectares of land available for housing, giving a capacity for just over 15,000 new dwellings at an average net residential density of 33.7 dwellings per hectare. The main areas for housebuilding were to be on some large peripheral sites such as Croxteth Park and land at Fazakerley Hospital and on land provided by the MDC. Although there is no mention of housing in the city centre, some progress had been made in getting private housebuilders to invest in the inner areas. Since 1976 the Council had pursued a policy of 'building for sale' to encourage the provision of low cost private housing. This had involved the council entering into agreements with private developers to build houses for sale on council owned land. This ground-breaking scheme had proved very popular and by 1983 just over 2,000 dwellings had been provided on 27 inner area sites. The effect of the scheme was to add significantly to the city's private housing stock and provide increased scope for residents to filter up through the local housing market. A further and in the longer run, more significant, effect was that the scheme was one of the first in the country to demonstrate that private housebuilders would be prepared to invest in the inner areas under the right economic conditions. This discovery was to provide one of the foundations for a dramatic shift in policy that would encourage private housebuilding as a tool for regenerating inner urban area over the next two decades (Couch and Wynne, 1986). Figure 7.1 below indicates the location of these and other major housing initiatives in the city since 1976.

Another objective was the creation of an urban environment with its own identity and character: healthy, safe, well-maintained and conserving the best features of the past. Little was said about how this environmental strategy was to be implemented, neither did the document present any analysis of the existing townscape nor any strategy for urban design. The existing environment was to be improved through a number of specific proposals (such as a riverside walk along the Mersey) and two area-based policies: Industrial and Commercial Improvement Areas and Environmental Improvement Areas. However, the funds allocated to these areas were so meagre that only the most modest improvements could be anticipated. Rather more attention was devoted to protecting the city's built

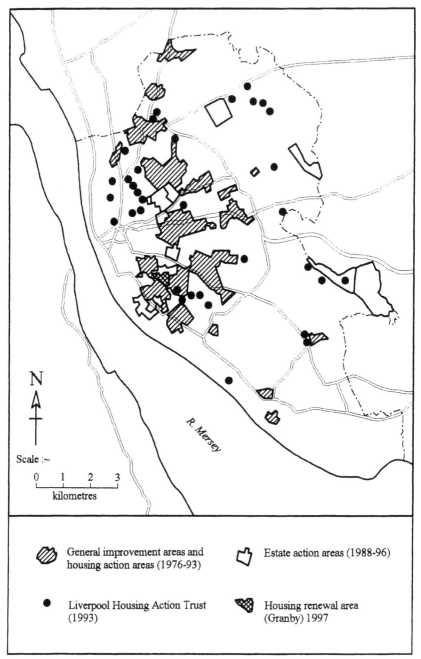

Figure 7.1   Major housing regeneration initiatives since 1976
Source: Liverpool City Council 1983, 1996

heritage. Over the previous decade more than twenty additional conservation areas had been designated within the city, bringing the total to 30 areas covering 838 hectares (about 7.5 per cent of the total area of the city). These additional conservation areas extended coverage of the city centre around Dale Street and Old Hall Street; Newsham Park; village centres in Wavertree, West Derby and Woolton; and innovative residential developments from the 20th century such as Wavertree Garden Suburb, Mill Bank and Muirhead Avenue. The weakness of the local economy had benefits for conservation in that the demand to modify or demolish historic buildings was probably lower than in many other more prosperous cities. On the other hand and for the same reason, it was difficult to attract private investment into building restoration or conversion projects.

Despite the massive decline in population and the changes in retailing that had occurred during the previous two decades the Council's policies for the city centre and the ten designated district centres remained largely unchanged. The document anticipated 'no major structural changes in suburban retailing centres' and yet the 1980s were to see some of the most significant changes in retailing since the Second World War. What followed was a rapid growth of freestanding superstores, discount warehouses and large specialist DIY, furniture and electrical goods outlets, all serviced by large free car parks. Whilst these developments had only a limited impact on the city centre, local shops and district centres suffered a considerable loss of trade. The problem of declining demand for small shops and increasingly obsolescent shopping streets was acknowledged and policies put in place to improve the shops themselves and to enhance the local environment. However, this policy of marginal improvements was of little benefit in a situation where few of these small local shops could ever compete in the new retail economy. Although there was some demolition of obsolete shops the reality was that the city had, and still has, far too many shop units. Policies designed to protect local and district centres from competition were to prove of little long-term benefit to their local environment or to the economy of the city.

Policies for recreation were confined to open space, allotments and indoor recreation. The approach to public open space was primarily quantitative, comparing the city's provision favourably against national standards but noting deficiencies in certain parts of the inner areas. However, the quality of some of the city's existing parks was giving cause for concern as they continued 'to exhibit problems associated with inadequate maintenance and vandalism', which had made many facilities, such as changing rooms, virtually unusable and deterred many people from visiting parks. It was proposed to use money from the Inner Area

Partnership to upgrade some of these parks, although the amount of money available was woefully inadequate for the task. At the same time and with questionable wisdom, the City Council was providing a considerable amount of new public open space on former slum clearance sites within the inner areas. Whilst responding to the perceived quantitative shortage through capital investment in 28 hectares of new open space, funded through the Urban Programme, the Council had insufficient funds to provide the revenue support necessary for the adequate management and maintenance of these spaces. In consequence much of what was created was of little aesthetic or recreational value.

There was also a shortage of indoor recreation facilities and swimming pools. According to national standards the city should have had 13 purpose-built sports centres for public use, whereas it actually had 5 available to the general public together with a further 5 primarily for school use. Again it was anticipated that money would be made available from the Inner Area Partnership to fund up to 3 new sports centres, together with the conversion of the MDC's Festival Hall into a major new leisure centre after the end of the International Garden Festival.

By this time the City Council had little control over transport policy, which was now mainly the responsibility of the Merseyside County Council. Much of the chapter on transport was little more than repetition and support for Merseyside Structure Plan policies in this field, although concern was expressed about the inadequacy of expenditure on highway maintenance. Thirty years earlier the Buchanan Report had seen car parking as an effective tool to control traffic levels, yet in a reversal of earlier policy, in 1983 the City Council's concern was to increase the accessibility of the city centre to motorised traffic:

> With a continued rise in the level of car ownership and usage, the provision of adequate and convenient car parking facilities...is essential if existing shopping and commercial centres are to maintain their attraction. This is particularly true of Liverpool City Centre (Liverpool City Planning Department, 1983, para.9.27).

Clearly the political imperative at this time was to protect the economy of the city centre before considering the impact of traffic and its control. But within four years the City Council did undertake a more wide-ranging review of its planning for the city centre.

Liverpool City Centre: Strategy Review, 1987

Little consideration had been given to the idea of a comprehensive plan for the city centre since the reorganisation of local government in 1974 but by the mid-eighties there was renewed interest in the area. Manchester City Council had produced an innovative local plan for its city centre in 1984 that recognised the importance and potential of the city centre to the economy of the modern conurbation. In 1987 Liverpool City Council follow suit with a document that:

> should provide the starting point for a comprehensive re-evaluation of planning strategy in the city centre which has remained largely unaltered since the 1965 City Centre Plan (Liverpool City Council, 1987, p2).

The review identified a number of problems with the city centre, including its environment, security, pedestrian and vehicular circulation, but also recognised the importance of the centre to the economic, social and cultural life of the conurbation. A number of themes were developed.

The first was to respond to changing retail market trends by seeking recognition and support for Liverpool's special role as a regional shopping centre. The second theme was the need to strengthen links between Liverpool and its catchment area. This was a call for more investment in highways and public transport services connecting the centre with its hinterland. Controversially this included calling for the abandonment of the Mersey tunnel tolls (Liverpool City Council, 1987, p10). The third theme was improving the circulation system for pedestrians and vehicles within the centre. Here emphasis was put on the 'legibility' of the circulatory system, and on improving arrival points, such as bus and railway stations, and improvements to the pedestrian environment. Theme four was concerned with upgrading the environment through continued protection and enhancement of the historic building stock, investment in the public realm and stronger development control policies.

Theme five was much more novel, recognising the importance of tourism and its potential to boost the local economy, it was proposed to develop a tourism plan for the area. Theme six considered changing commercial requirements and the need to upgrade much of the outworn office floorspace. Theme seven introduced the notion of increasing housing opportunities in the city centre, although far from being based upon the view that a return to urban living would be environmentally and culturally beneficial, the argument was much more prosaic. Housing developments would facilitate the sale of land, re-use of vacant buildings, construction employment and increase the Council's financial income.

Reinforcement of the centre's artistic and cultural importance was the eighth theme.   This was based upon an argument for cultural-led regeneration, also a new idea at the time. The final theme was to promote a positive image through marketing, lobbying and use of the media.

Many of these themes were very differed from the issues that had concerned the 1965 plan. There the emphasis had been on the physical modernisation of the centre, with great ambitions for new infrastructure and replacing much of the building stock. Here much less emphasis was placed on large-scale investment. Physical improvements was to be achieved through conservation of the built heritage and piecemeal improvements elsewhere. Importantly there was a discussion about the role and function of the city centre in the modern economy and an understanding of the importance of considering new issues: city centre living, culture, tourism and image. The review was a key document in shaping the debate and research that led to the production of a new City Centre Plan in 1993, which was in turn incorporated into the emerging Unitary Development Plan. The debate also helped shape the policies of other agencies through the 1990s and many of the ideas that appeared to be novel at the time subsequently became accepted elements of mainstream urban policy.

The Liverpool Unitary Development Plan, 1996[1]

Merseyside County Council was abolished under the Local Government Act 1985 and new arrangements were made for strategic planning in metropolitan areas.   Each metropolitan district was required to prepare a 'Unitary Development Plan' (UDP) that contained both general (strategic) policies as well as detailed site-specific proposals.   Thus Knowsley, Liverpool, St Helens, Sefton and Wirral councils were each required to prepare Unitary Development Plans for their own areas.   Co-ordination would be provided through the Department of the Environment.

By the 1990s Planning Policy Guidance Notes (PPGs) had become an important mechanism used by central government to steer the nature and content of development plans.   From 1988 guidance notes had been prepared on a range of planning matters including Housing (PPG3), Town Centres and Retail Development (PPG6), Transport (PPG13) and the Historic Environment (PPG15).   Within Merseyside the area-specific PPG11, 'Strategic Planning Guidance for Merseyside' (Department of the Environment, 1988), provided the guidance necessary to co-ordinate the plans of the metropolitan district councils with regard to economic

---

[1] All discussion of the Liverpool UDP in this book refers to the 1996 Draft.

development, housing development targets, the green belt, recreation and tourism, major shopping proposals, waste disposal and transport. Furthermore, whereas upon its establishment in 1981 the MDC had been required to take account of the Merseyside Structure Plan in formulating its proposals, in a reversal of power, the Liverpool UDP was now expected to have regard to the policies and proposals of the MDC. Thus through policy guidance and the influence direct investment central government and its agencies were imposing more constraint on development plans than ever before.

A Consulation Draft of the Liverpool UDP was produced in September 1994 with a 'Deposit Draft' being published two years later.[2] The UDP would replace the Liverpool Development Plan (1958), the Merseyside Structure Plan (1980) and the Merseyside Green Belt Local Plan (1983) as the statutory development plan for the city.

In his forward to the Liverpool UDP Councillor Frank Prendergast, Leader of the Council, set a very different tone from that of Alderman Sefton two decades earlier. Although there is still an ambitious desire to create 'a premier city for the 21st century', the heroic ambition to rebuild the city is gone. In its place is a more pragmatic objective concerned to make the city an attractive proposition for potential investment:

> Liverpool is once again establishing itself as a vibrant and dynamic city for its people to be proud of. In order to establish the future of Liverpool as a premier city in the 21st century, we need to provide the right framework to guide development, to ensure that the city is an attractive place in which people will want to live, work and invest as well as visit (Liverpool City Council, 1996, p5).

The major themes of the UDP were economic regeneration, environmental improvement and reduction of inequality. Whilst the first two themes could be said to reflect advances and developments in national planning policy through the preceding decades, the third was perhaps more of a local than a national political concern at that time. Part one of the UDP set out the strategy for the development of the city, whilst part two comprised more detailed development control policies and specific land use proposals. After the introduction and provision of some contextual information the strategic objectives and policies were expressed in a single chapter (Ch 5) comprising 11 pages. By contrast part two required 259 pages and nine chapters to set out detailed policies for: economic

---

[2] Despite publication in 1996 and a lengthy public inquiry the final version of the UDP had still not been approved by the millennium.

regeneration; heritage and design in the built environment; open environment; housing; shopping; transport; community facilities; environmental protection and Liverpool city centre.

The context within which the plan was devised was depressingly familiar. The city had lost a further 60,000 people between 1981 and 1991. At least this did indicate a slowing of the rate of population loss over the previous decade. Whilst slum clearance and overspill were no longer a major cause of out-migration, many more affluent people were still leaving in order to improve their housing conditions or for job-related reasons. The unemployment rate within the city remained at about twice the national average; 41 per cent of households in Liverpool lived in poverty and 16 per cent in 'intense poverty' and despite the achievements of the renovation programme, nearly 200,000 dwellings were considered unsatisfactory. (Liverpool City Council, 1996, pp25/26).

A novel feature of the Liverpool UDP at the time was the use of an 'environmental appraisal' to assess the extent to which its policies took account of environmental issues including global sustainability, control of natural resources and local environmental quality. The Liverpool UDP was one of the first such plans to respond to the new national requirement to produce such appraisals. Another feature of the UDP was the attempt to place the document within the framework of the City Council's corporate policy. However, compared with the days when it was a county borough council, the local authority's powers had been emasculated and unlike its pre-1974 predecessors, the plan did not attempt to include the spending programmes of service departments. Although in 1996, the City Council had control over its own housing, highways, education provision, social services, parks and amenities, its own capital budgets were small. Most housing was by this time provided by the private sector with a small additional contribution of social housing being made by independent housing associations (registered social landlords). Local authority house building had all but ceased. Control over further and higher education had been lost and a growing number of schools were being grant maintained directly by central government. Public transport investment depended upon Merseytravel (the passenger transport authority) and an array of private suppliers. Much local investment in regeneration projects was being undertaken by short-life ad-hoc regeneration agencies whose programmes were only loosely influenced by the corporate or development policies of the City Council.

The ambitious aim of reversing the decline in economic activity was to be achieved through a policy of concentrating resources in five regeneration areas: the City Centre; Waterfront, Docks and Hinterland (approximating to the former MDC area); the Eastern Corridor (including Wavertree Technology Park); Speke/Garston; Gillmoss/Fazakerley and Aintree. With the exception of the provision and servicing of sites for economic development, the remaining strategic policies for economic regeneration amount to little more than a statement of aspirations: 'encouraging small-scale economic development', 'strengthening the commercial role of the city centre', 'promoting and enhancing the role of the airport and docks', 'promoting the principle of mixed-use', 'maximising the contribution of telematics to economic regeneration', and 'promoting the role of Liverpool as a regional cultural and tourism centre'.

The primary source of land for economic development would be found from the recycling of vacant and derelict premises whilst the city's green spaces and peripheral areas would be protected from unsuitable development. During the nineties the notion of 'mixed-use' developments became strongly endorsed as part of national planning policy and was promoted by the Liverpool UDP on the basis that it would 'help to reduce the number of journeys made by private car'. The UDP designated a number of 'mixed-use areas' and 'sites for various types of development'. A related policy 'the 24-hour city' was also embraced by the UDP through promotion of the city centre as a lively and exciting place, and by providing a range of facilities throughout the day and night. Most of the detailed policies for economic development were statements indicating how the authority would respond to applications for planning permission rather than commitments to any investment by the City Council itself, very different approach from the plans of the 1960s and 1970s.

Much of the chapter on Heritage and Design was concerned with development control policies for the protection of the rich heritage of listed buildings, churches, parks, gardens, trees, cemeteries, ancient monuments, archaeological remains, and conservation areas. Prior to the UDP the city possessed 32 conservation areas; one (Mill Bank) was now to be de-designated, five new ones created and the boundaries of a number of others adjusted. There was a statement of fourteen 'general design requirements' against which proposals for new development would be assessed. These included, for example: requirements that the development provided an attractive façade using external materials of lasting quality sympathetic to its location; that the bulk and height of new buildings be of a scale which would complement the surrounding area; and that the building lines and layout of the development would relate to those of the locality. In the 1960s Liverpool's planners had devised maps showing where tall buildings of various heights would be permitted or prohibited. The UDP confined itself to a statement that views of important landmarks contribute to the character of the city and that tall buildings can block or spoil these views. Further policies dealt with the provision of access for the disabled; using the design of the physical environment to reduce the opportunities for crime; energy conservation; the planting of trees; provision of public art; the control of advertisements, telecommunications installations and light pollution.

Policies for 'the open environment' would replace the Merseyside Green Belt Local Plan (1983) within the city. In addition protection would be provided for two 'green wedges' (Calderstones/Woolton and Otterspool) that brought open land into the heart of the city. There were also policies to protect the Mersey coastal zones; nature conservation sites and features. The provision of parks and recreational open space continued to be a concern but by 1996 the population of the city had fallen to such an extent that the target figure of 2.4 hectares per 1,000 people had now been achieved by default. Of greater concern was the poor quality and management of much of the city's open spaces. A 'parks strategy' was being prepared.

In support of economic development a programme was put forward for environmental improvement corridors along the major routes into the city. There were also proposals for the creation and enhancement of natural habitats, for example through the protection of wildlife corridors, better management of open land and planting the 'Mersey Forest'. This last initiative was the City Council's contribution to the Countryside Commission's proposal to create 8,140 hectares of new woodland on Merseyside under its Community Forest Programme.

Although many of the heritage and environment policies appeared to be relatively weak and hedged with phrases such as 'the City Council will encourage', they did represent a widening of the planning agenda and a concern for the quality of life of local people that had not been present in earlier plans.

By the mid 1990s the provision of additional dwellings was no longer a significant political issue. Nevertheless, responding to government guidance, the UDP made provision for a modest addition of around 23,100 dwellings in the period between 1986 and 2001. Of greater concern was housing renewal. In the public sector an 'Estate Action' programme during the early 1990s provided the main vehicle for efforts to improve the quality of existing council estates. But the problems were considerable and in totality, their solution lay beyond the resources of the City Council. Out of 48,000 council owned dwellings, 3,400 were classified as unfit whilst a further 32,000 were identified as requiring renovation (Liverpool City Council, 1996, p192). In addition to the Estate Action programme, in order to tackle one specific and very severe problem: that of multi-storey flats, the City Council had reached an agreement with central government to establish a Housing Action Trust (HAT) to address the problems of the city's tower blocks. In all some 67 blocks were transferred to the Liverpool HAT in 1993 (see below). Other policies included the 'Vacant Dwellings Initiative' which brought together the City Council, local housing associations and private developers to address the problems of 'difficult to let' estates. In one example, St Andrew's Gardens a former inner city council estate comprising inter-war walk-up flats and post-war maisonettes was transformed through renovation and selective demolition and replacement to provide a mix of student accommodation, private and social housing.

Figure 7.2   **St Andrew's Gardens**
Former council flats converted into student accommodation; maisonettes down-topped to form houses; and new housing fronting onto Russell Street. A number of tower blocks that can be seen in the background have since been demolished by the LHAT

Source: author

In the treatment of obsolete private housing, the long-standing programme of GIAs and HAAs had proved effective in improving more than 28,000 dwellings since the mid-1970s. However, there had been little progress since 1989 when a change in legislation effectively brought the programme to a halt in Liverpool. A new grant regime changed the funding mechanism from one based upon the cost of renovation to one that was limited by the income of the householder (means tested). Under this new legislation the City Council had eventually declared a Renewal Area covering more than 700 dwellings in the Granby area (where Shelter had undertaken their SNAP action-research project 20 years earlier), but this was modest beer compared with the previous level of activity. Although the UDP says surprisingly little about the condition of private housing stock this was an issue that would become a major concern before the end of the century.

City centre living was also being promoted by the UDP through the designation of zones within the centre as suitable for residential developments and by providing a development control regime favourable to the conversion and development of property for housing. By the mid-1990s real progress was being made in encouraging such investment (Couch, 1999).

A priority of the UDP was to concentrate retail provision into existing shopping centres, although limited out-of centre developments were to be considered favourably under certain circumstances. Despite the dramatic changes in the retail economy over the previous forty years, the hierarchy of existing shopping centres recognised by the plan differed little from that of the IPPS back in 1965. Liverpool city centre was expected to remain the main location for comparison shopping even though it had lost much of its regional dominance to competing centres in Chester, Warrington and elsewhere.

London Road, an outlier of the city centre, had been declining for many years. However, it had been included within the Liverpool City Centre East City Challenge programme, and a London Road Development Agency had been established In consequence the area was by now benefiting from a pedestrianisation scheme, landscaping and other improvements to the public realm that were in turn bringing about a consolidation of its shopping function and new investment in office and residential floorspace (see below).

The City Council still clung to the concept of 'district centres' that would provide an intermediate level of service between the city centre and local shops. Some 12 such centres were designated – virtually the same list that had appeared in the 1965 plan. Unfortunately, as a consequence of declining population, rising car ownership and changing patterns of shopping, many of these district centres were suffering from high levels of vacancy and a lack of investment. The UDP proposed various improvements to district centres but without strong support from the retail sector it was unclear how such projects were to be funded or implemented. Similarly many local shopping centres and individual local shops were by the mid-nineties suffering a similar fate. Investment was low and vacancy rates rising. Nevertheless the UDP sought:

> to maintain and enhance the role of local and neighbourhood shopping centres, in order that they may provide a wide range of convenience goods to serve local everyday needs (Liverpool City Council, 1996, p218).

Precisely how this was to be achieved was unclear. Shopping provision was in many ways one of the most problematic policy areas for the UDP. The aim was to preserve the existing hierarchy for the commendable reason of providing accessible shopping for all the community and to reduce the need for car journeys. But the economics of retailing were changing the historic pattern of retailing beyond recognition and despite the concerns of the planners, out-of-centre retail parks and superstores were opening apace. What is more, the public appeared to welcome these changes and voiced few objections. The point was that, just like to decline of manufacturing industry over the previous two decades, these changes were being driven by forces beyond the control of local people and beyond the control of individual retailers who had no choice but to follow the trends. In such circumstances the opportunity for the UDP to protect the existing shopping hierarchy from change was minimal.

Transport had been the first policy chapter to be considered in the IPPS, where it was seen as the glue that bound together the urban structure of the city. Thirty years later in the UDP transport policy was to be given nowhere near the same prominence. This was not because of any diminishing concern about the importance of transport policy to the future of the city but because of the very limited scope for the UDP to 'do anything about it'. The transport strategy contained within the UDP had been developed from the 1992 Merseyside Integrated Transport Study (MERITS). The main thrust of this strategy was based upon maximising the performance of the existing transport system and promoting the role of passenger transport, cycling and walking in meeting the city's transport needs. To this end the UDP proposed new or improved bus facilities in the city centre and traffic management measures designed to enhance the speed and reliability of bus journeys, although initially only one radial route (A59, County Road) was identified for improvements. New railway stations were proposed at Brunswick and Wavertree Technology Park with various improvements at other stations. However, rail investment was the responsibility of Merseytravel and the railway companies. Here the UDP was not so much planning rail investment as reporting the plans of others. There were proposals for studies into the feasibility of rapid transit systems on radial routes to Page Moss, Croxteth and Netherley. Cycling was to be encouraged through the adoption of a 'Cycling Strategy for Liverpool' which included more and better cycle routes, better signage, road safety measures and secure cycle parking. Walking and the pedestrian environment were also to be made safer and more convenient by a variety of measures.

Despite these commendable commitments to an environmentally sustainable transport policy, some new road schemes were proposed, most controversially the New Russell Street (Russell Street-Berry Street link) which was to have been funded through the City Challenge programme and which was later thrown out by central government. The other major road investments proposed for the city were the Sandhills Link, to provide an east-west link between Scotland Road and Regent Road/Great Howard Street; and the Low Hill Distributor, to act as a north-south distributor to the east of the city centre. Whereas road building had been seen as a key object of urban policy in the 1965 plan it was now expected that with these exceptions, traffic would be accommodated within the existing road system. In 1965 the provision of city centre car parking had been related to and limited by the capacity of the primary road system but in 1996 the UDP proposed simply to 'ensure that adequate parking facilities are provided to meet the needs of residents, shoppers and visitors' (Liverpool City Council, 1996, p267). However, proposals for major developments throughout the city were to be subject to 'traffic impact assessments'.

As with the environmental policies a number of these transport policies, especially those concerned with the promotion of public transport, cycling and walking, appeared to be relatively weak. The City Council had limited power to take policies through to implementation and there were real concerns about the absence of co-ordinated planning within the transport system generally. Nevertheless, as with environmental policy, transport policy attempted to widen the planning agenda and there was recognition of the importance of accessibility rather than mobility and a concern for cycling and walking that had been absent from earlier plans.

One of the strategic policies of the UDP was to ensure the satisfactory provision and distribution of community facilities. Detailed policies were elaborated for social facilities (such as churches and community centres); care facilities (residential care homes, day centres, child care); health care (including hospitals, health centres and GP services); education facilities, from schools to universities and colleges; Liverpool and Everton football clubs; indoor and outdoor sports facilities and play areas. In truth the City

Council had very little control over the provision of most of this collection of facilities, which in fact had very little in common with each other to justify being brought together in this way. Policies were mainly aspirational (the City Council will support...) or reactive (concerned with providing a framework for development control). As with so much else in the UDP (housing, public transport, shopping) there was little sense that City Council was planning the future provision of community facilities. For the most part it was reacting to the plans and policies of others.

The City Council encouraged the reclamation and reuse of derelict land, giving particular priority to projects that would benefit economic regeneration or that were in a particularly hazardous state. But again, most of the initiative would have come from other agencies such as English Partnerships or the MDC, each of which would have their own plans and priorities for site development. Various detailed policies were set out for the control of development in relation to landfill, landfill gas, waste reception, storage, disposal and recycling. The formulation of a Waste Disposal Plan had, since 1986, been the responsibility of the Merseyside Waste Disposal Authority but passed to the new Environment Agency in 1996. There were also detailed policies for the control of pollution. In another example of widening the planning agenda without any obvious ability to deliver, the City Council would 'support the development of renewable energy projects provided that the facility is in accordance with other policies in the Plan' (Liverpool City Council, 1996, p316).

The final chapter brought together policies for the city centre. The chapter was derived from the 1987 City Centre Strategy Review and the 1993 City Centre Plan. The strategic objectives for the city centre were seen as meeting the needs of the community, having a good physical environment, and providing space for development. Much of the chapter was little more than a repetition of policies for the city centre that had been set out elsewhere in the UDP with regard to retailing, office development, community facilities, arts and culture, and transport.

Conclusions

Within the City Council plan making in the seventies had been frustrated by slow progress on the structure plan and internal political difficulties. However, in 1983 a statement of current planning policies was published. By this time economic development, urban regeneration and environmental improvement had become the dominant themes of planning in Liverpool. Transport planning had slipped down the agenda and there was little

planning interest in the city centre until later in the 1980s when the Council
began a dialogue about its future role and development.

**Table 7.1   The nature of planning and urban regeneration in
Liverpool during the period of a return to comprehensive
planning and widening the regeneration agenda**

What was the relationship between planning and urban regeneration?	Urban regeneration was setting the local policy agenda. Development plans were increasingly expected to accommodate rather than formulate urban regeneration policies.
What were the main achievements of planning and regeneration during the period?	Many 'brownfield' sites and 'heritage' building re-used. Evidence of an emerging re-urbanisation process with growing investment and a return of population to the city centre.
What was the relationship between economic, environmental and social aims?	Growing environmental dimension to policy making. As the economic situation eased there emerged an improving balance between economic, environmental and social aims.
What was the extent of local democracy and participation?	Central government continued to exercise tight control over local government and set the planning agenda. However, the role of local government clarified as facilitator and co-ordinator of urban regeneration programmes. Continued use of non-democratic agencies and the private sector to implement urban regeneration programmes. Strong encouragement for community participation in urban regeneration at the local level.
To what extent were policies co-ordinated as part of long-term strategy of fragmented and short-term?	Rapid changes in the institutional and policy framework for planning and urban regeneration. Despite rhetoric to the contrary both central and local government policies appeared to be fragmented and lacking in co-ordination. Local planning shifted towards a satisficing (criteria based), rather than optimising (zoning) approach to the control of development.

Source: author

One of the most striking features of urban regeneration at this time
was the domination of area-based policies.   As well as major area
designations such as the Inner City Partnership Area and the designated
areas of the MDC and Speke EZ, there were Industrial and Commercial
Improvement Areas, over a hundred HAA and GIA, more than thirty

Conservation Areas and a number of Environmental Improvement Areas. The argument for this focus on specific areas was that it permitted a concentration of resources that would lead to better outcomes than if those same resources had been spread across the whole city.

As indicated in Table 7.1 above, the Unitary Development Plan was very different in nature from earlier development plans. Responsibility for so much policy making had, by this time, passed to other agencies that it became far more of a document reporting the policies and proposals of others than being the source of original proposals itself. Nevertheless, the UDP was interesting for the way in which it attempted to widen and modernise the planning agenda. The understanding of sustainable development and its sensitivity to environmental and social concerns was far greater than in previous documents. Its development control policies were much more sophisticated than earlier plans, and it was through these policies, rather than through large-scale investment proposals, that the UDP attempted to provide a framework for the future development of the city. Although it did little to initiate development and cannot even be said to be a very effective mechanism for policy co-ordination, it did provide a set of criteria to be met by any development proposals. It was in this way that the many of the aims of the UDP were to be achieved. But despite their shift in style development plans remained too narrow a policy instrument to deal with all the problems of urban change and achieve the City Council's ambitions for economic development, environmental sustainability and social inclusion. To achieve this there would have to be a widening of the regeneration agenda also.

8 Widening the Regeneration Agenda

The 1990s saw significant changes in the approach to urban regeneration in the city. While there was progress in widening the agenda there was also increasing fragmentation and difficulties in co-ordinating the many players that were now involved with the regeneration process. Despite the marginalisation of local authorities that had occurred under Thatcherism, it had become clear, even to the Conservative government, that such a position was untenable. The first signs of change came with the Local Government and Housing Act 1989. That Act created Renewal Areas and encouraged local authorities to take a leading role in their running. The second came with Michael Hessletine's City Challenge initiative. Housing Action Trusts were another idea that could only be implemented after the policy had been modified to include a role for local authorities. On its re-election to power in 1993 the Conservative government recognised a need for a substantial review of the government machinery surrounding regeneration policy. Regional Offices of Government were established to improve inter-departmental co-ordination in the delivery of services. English Partnerships was established as a regeneration agency for England and funding was reorganised into a Single Regeneration Budget. The same year, 1993, also saw Merseyside designated by the European Union under 'Objective One', giving the area access to substantially increased European funding for economic development. There were changes to the strategic planning framework affecting Liverpool. The County Council had long since gone but a new regional tier of governance was emerging: initially through the preparation of informal proposals for regional development and after the return of a Labour government in 1997, the establishment of the North West Regional Development Agency and a Regional Assembly. Finally, towards the millennium, the city benefited from a number of new government's area-based regeneration initiatives including a New Deal for Communities programme in the Kensington district and an urban regeneration company: Liverpool Vision, in the city centre. This chapter considers the widening of the urban regeneration policy agenda under these various initiatives.

Renewal Areas: Granby

Renewal Areas were introduced in 1989 to encourage co-ordinated action to improve housing, environmental, social and economic conditions in run-down neighbourhoods. They were aimed primarily at districts containing poorer quality private housing occupied by lower income households. A particular concern of government was their view that recent housing renewal activity had put too much emphasis on housing improvement and failed to give sufficient consideration to related local social and environmental issues and local economic development. In establishing a renewal area, local authorities were required to prepare a comprehensive analysis of problems and potential solutions through a 'neighbourhood renewal assessment' undertaken in collaboration with local stakeholders, including residents and other users of the area. This marked a departure from previous more 'top-down' approaches and represented an attempt to tackle problems at a grass-roots level as well as encouraging corporate working within local authorities and more effective multi-agency co-ordination (Couch, Eva and Lipscombe, 2000).

Liverpool itself was slow to take up the opportunity offered by this new approach. However, a renewal area was eventually declared in Granby in 1996, where SNAP had been 25 years earlier. Following detailed consultations with the Granby Residents Association and the wider community, proposals for renewal of the area were agreed. In this multi-cultural area the main objectives of the programme were for local economic development, housing renewal and environmental improvements, all to be developed within a framework of community involvement. The situation in Granby at that time was that 32 per cent of dwellings were potentially below national habitation standards and a further 56 per cent in need of general or substantial repairs. Environmental problems were exacerbated by the presence of empty or derelict houses, the condition of back alleyways, rubbish, litter and poor street lighting. Household incomes were low with 70 per cent of residents having incomes below £7,500 pa (Liverpool City Council, 1995). Proposals included group repair and housing improvement schemes, clearance of a considerable number of dwellings, building new homes on the sites that would become available, and environmental works to improve boundary walls, alleyways, pavements and roads. By the millennium much of this work had been completed with major improvements to the physical environment (see Derbyshire A, 2001, for a wider account of regeneration in Granby).

Through Renewal Areas the government had established a new approach to neighbourhood renewal that was: holistic (considering environmental, social and economic issues); and inclusive (facilitated by the local authority but involving the participation of local residents and other stakeholders in the area). The new Labour government would later take up this approach in its national strategy for neighbourhood renewal.

Figure 8.1   New housing in the Granby Street Renewal Area
As a result of the Renewal Area much of the former 19th century terraced housing has been demolished and replaced. Little is left of the once-thriving, multi-cultural shopping street

City Challenge: Liverpool City Centre East

Following his return in 1990 to the office of Secretary of State for the Environment, Michael Heseltine announced a new urban regeneration initiative: City Challenge. Under this programme the Department of the Environment invited bids from the most deprived local authorities for proposals to spend £37.5 million each over a five-year period. Each bid was to contain a clear vision for regeneration supported by a worked out strategy for its realisation. Bidding was to be led by local authorities but proposals had to be prepared in collaboration with other local partners from the business, voluntary and community sectors. City Challenge was innovative in that it was the start of a return to a recognition by central innovative in that it was the start of a return to a recognition by central government of the key role that could be played by local authorities in facilitating development; it was overtly competitive (only 11 of the 15 initial bidders would be successful); it encouraged local partnerships and a participative approach to regeneration; and it forced local councils into thinking clearly about their aims for regeneration programmes (Couch, 2000, p158).

Liverpool's City Centre East was one of the successful bids in the first round of City Challenge funding. The proposal was self-evidently the result of partnership working as it was signed by more than 30 representatives from the public and private sectors. Those represented included the local authority, voluntary housing sector, central government agencies, the universities, the churches, various business organisations and even Paul McCartney. Perhaps the only group not adequately represented in this partnership were local residents. The 'vision' set out in this bid document was:

> A vision of enhancement, activity, partnership, participation and quality of life. New jobs will be created in growth sectors – the arts and cultural industries, the visitor business, high-tech medical and further and higher education. The magnificent architecture and townscape will be conserved, enhanced and given a new life and purpose. Run-down areas will be regenerated and integrated with the mainstream of the city's life. It's a vision of physical regeneration, people, enterprise and growth sectors and effective and sustained management (Liverpool City Challenge, 1991, p6).

The so-called 'city centre east' was never a coherent or easily identified area for it included elements of the city centre in need of redevelopment (Queens Square), a marginal retailing zone (London Road), a 'cultural' quarter (Hope Street) and the Canning Street conservation area. Thus it was no surprise that this 'vision' was in reality a loose bundle of disparate regeneration proposals bound together by little more than geographical proximity. The eclectic nature of the bid was further evidenced by the long list of different proposals to be included within the five-year programme. These were presented under four headings:

- *The physical vision:* a new link road (completing the Liverpool inner ring road); a guided light transit system that could link the area with the rest of the city centre; the 'rebirth' of St George's Hall; rebuilding Queen Square 'with a quality of architecture and design in keeping with its magnificent setting'; bringing Lime Street Chambers and various London Road properties back into beneficial use; refurbishing and redeveloping St Andrew's Gardens to provide housing and student accommodation; removing through traffic from Rodney Street; restoring the Canning Street conservation area; refurbishing and re-use of major buildings in the Hope Street area (Blackburn House, Liverpool Institute and the Philharmonic Hall); a new urban square at the junction of Hope Street and Mount Street.

- *The people vision:* comprehensive and sustained training; increased employment opportunities for local people; increased provision to facilitate entry or re-entry into the labour market; improvements in the quality of life for local people through measures embracing increased housing choice, improved residential environments; better access to quality social, recreational, cultural and health facilities.
- *The enterprise and growth sector vision:* measures to encourage the development of small enterprises; partnerships with higher educational institutions spinning off small high-tech and consultancy businesses; maximising the enterprise potential of change in the health service; development of a vibrant craft, arts and cultural industries sector; increased visitor business based on enhancing the area's major attractions.
- *The management vision:* creation of innovative, locally based and effective management and maintenance; develop models of good practice; lay foundations for managing and maintaining the area beyond 1997; ensure the necessary level of dedicated effort from within the City Challenge partnership.

Most controversial was the proposal to complete the inner ring road (Russell Street – Berry Street link) that was claimed to be vital to the future of the area, only to be turned down by the Secretary of State some years later as being an unnecessary and old fashioned solution to the area's modest local traffic problems.   Nevertheless, the bid was successful and a 'partnership' agency was established to implement the programme. Progress was impressive and by the end of the decade much of the intended physical refurbishment and redevelopment was complete. This included:

- New public realm around Pembroke Place/London Road
- Refurbishment of the former Collegiate and SFX schools for new uses
- Works to St George's Hall

- Conversion of Lime Street Chambers (the former North Western Hotel) to student accommodation
- Redevelopment of Queen Square, including a new hotel, refurbished council offices, the Conservation Centre, a public square, bus station, shops and leisure facilities
- Regeneration of the St Andrew's Gardens area as student, social and private housing
- Refurbishment of derelict properties in Mount Pleasant for residential use
- Renovation and extension of the Philharmonic Hall
- Conversion of Blackburne House (a former school) into a Women's Education and Technology Centre
- Conversion of the former Liverpool Institute into the Liverpool Institute for the Performing Arts
- Enhancements to the Canning Street conservation area.

Looking at the general impact of City Challenge programmes nationally, the Department of the Environment concluded that they had brought a more flexible and innovative approach to regeneration and developed the partnership approach at the local level. Local authorities had streamlined their decision-making and programme delivery and progress had been made in developing more effective systems of project appraisal, management and monitoring (Department of the Environment, 1996, p178). Commenting specifically on the Liverpool City Centre East the DOE were complimentary, noting that there had been particular success in property development, bringing back into use buildings and sites that had been unoccupied for many years. It was also felt that the programme had acted as an effective lever for change in the area, improving image and investor confidence, and providing the foundations for continuing regeneration (Department of the Environment, 1996, p24).

Figure 8.2   **Canning Street**
City Challenge funding supported investment in this important
conservation area. Since the 1970s, when the area contained
many vacant and derelict properties, sensitive refurbishments
have turned this into a much sought-after residential location

Source: author

Liverpool Housing Action Trust

The idea of Housing Action Trusts (HAT) emerged in the 1988 Housing Act as a mechanism for renovating run-down council estates and removing them from local authority ownership and control. It had been the government's intention to designate HATs on a number of run-down estates of its choosing and to legitimise this action through referenda on the estates that gave tenants a choice between their present circumstances or the 'bright new future' promised through the HAT. Unfortunately the local communities in each case demonstrated their preference for the status quo (the devil they knew) rather than the government's vision, and all the proposed HATs were rejected. It seemed as if the concept had been stillborn until the passing of the 1993 Leasehold Reform, Housing and Urban Development Act which gave secure tenants a right to return to local authority tenure on completion of the renovation process. In these new circumstances six HAT proposals went to local referenda with local authority support and were accepted by the communities affected. One of the new-style HATS was proposed for Liverpool.

Unlike other Housing Action Trusts, which were area-based renovation programmes on particular run-down council housing estates, the Liverpool Housing Action Trust (LHAT) was unusually based upon a particular housing type: mainly high-rise blocks, scattered across the city. With the support of Liverpool City Council and local residents, LHAT was designated in 1993 (83 per cent of tenants had voted to transfer their properties to the LHAT). It took over responsibility for 67 of the city's 71 multi-storey blocks of flats, a total of 5337 dwellings.

LHAT's primary aim was to improve the quality of life for residents through programmes of enhanced management and renovation. Initially it was anticipated that the majority of blocks would be retained and refurbished. However, as the LHAT came to examine each of the blocks in detail it emerged that the costs of refurbishment had been grossly underestimated. By 1996 the original estimates for the cost of the works had more than doubled to over £350 million (Liverpool Housing Action Trust, 1998). A further problem was the declining demand for the occupation of tower blocks, even if refurbished. On designation,14 per cent of the flats were already vacant and it became clear that demand for all 5,337 properties would almost certainly never materialise in the foreseeable future however well refurbished. Further, more than 60 per cent of the existing tenants at that time were over 60 years old and 45 per cent of households contained someone with a disability. By 1999 LHAT had

Figure 8.3   The demolition of an obsolete tower block
Under the LHAT programme many of the city's multi-storey
blocks were demolished and replaced with low-rise housing
similar to that seen in this picture

Source: author

decided to demolish 44 blocks and to retain and refurbish only 16 blocks, with the future of 7 blocks still undecided at that time. In total only 3,350 flats were to be improved or replaced (Liverpool Housing Action Trust, 1999). As a consequence of these decisions LHAT became actively involved in the design and development of replacement housing for tenants. By March 1999, LHAT had, in partnership with other bodies, 933 new dwellings completed or under construction compared with the completed refurbishment for long-term use of only 132 dwellings (Liverpool Housing Action Trust, 1999).

Community participation was a major feature of the renovation and redevelopment programmes undertaken by LHAT. This included the initial referendum on the transfer of properties to the HAT, tenant members on the Board, the involvement of tenants groups in the design and development process, and tenant satisfaction surveys. Tenants were encouraged to participate in the design and redevelopment process through Project Advisory Groups (PAG) (later replaced by Neighbourhood Panels). These were established for each site and comprised tenants representatives meeting regularly with the professional team as the project progressed. Because, unlike the other HATs, the work of LHAT was spread across the city, there was no one single locational focus for regeneration activity and it was necessary to set up separate PAGs for each group of blocks. (For a fuller discussion of the LHAT and community based housing development see Couch, 2001).

Changing Directions in Urban Regeneration

By 1990 there was growing criticism that the government's urban regeneration policies were too many and too complex with too little co-ordination and too little recognition of the potential contribution of local authorities. The Audit Commission had commented, only a year after the publication of 'Action for Cities', that:

> Local government has an important role to play in (urban regeneration) ...local authorities believe that their role is undervalued by central government. They see themselves as increasingly marginalized. Government support programmes are seen as a patchwork quilt of complexity and idiosyncrasy. They baffle local authorities and business alike. The rules of the game seem over-complex and sometimes capricious. They encourage compartmentalised policy approaches rather than a coherent strategy (Audit Commission, 1989, p1).

This sober analysis of the Thatcherite approach to urban regeneration was particularly damning as it came not from academic or political sources but from one of the government's own agencies. The argument was that the whole was less than the sum of the parts because of the lack of any holistic view and with too many agencies involved, conflicting strands of policy brought confusion and diseconomies to the regeneration process.

At first the government did little to disentangle this web of policies and there was the 'diversion' of City Challenge brought in by Michael Heseltine. But following the appointment of John Gummer as Secretary of State for the Environment in John Major's government, 1993 saw the most radical shake-up of regeneration policy since the Conservatives had come to power in 1979. There were three strands to these changes:

- setting up Regional Offices of Government
- creating an English urban regeneration agency (to be known as English Partnerships)
- establishing a Single Regeneration Budget

The new Regional Offices of Government were intended to 'strengthen the machinery for co-ordination in the regions' and brought together the regional offices in England of the Departments of Trade and Industry; Employment; Environment and Transport. There were to be ten regional offices including the separate regions of Merseyside and North West England. The establishment of the regional office marked the end of the Merseyside Task Force.

English Partnerships (EP) was established in response to calls for greater co-ordination of regeneration spending and in order to complement the perceived successes of the Welsh and Scottish Development Agencies. EP brought together English Estates (the government's industrial estate developers), Derelict Land Grant and City Grant. The new organisation was to promote regeneration through the reclamation, development or redevelopment of land and buildings. It had power to financially support private development activity, undertake joint ventures and acquire and manage property. EP was essentially concerned with bringing land and property into beneficial use and less concerned with the wider social or environmental benefits of development. Despite many successes, EP was criticised for this strong orientation towards property development (Couch, 2000, pp164-167).

The Single Regeneration Budget brought together twenty-one different grants and spending programmes from five separate government departments into one single budget on the basis that funding would go where it was most needed locally rather than according to a set of priorities determined in Whitehall. In fact, more than three quarters of the budget came from the Department of the Environment and included the previously separate programmes of Estate Action, Housing Action Trusts, Urban Programme, City Challenge, Urban Development Corporations, Inner City Task Forces, City Action Teams, and the new Urban Regeneration Agency.

The purpose of the Single Regeneration Budget was to support regeneration and development that met objectives similar to those of the pre-existing programmes, *viz:* economic development, job creation, improving environmental and housing conditions, tackling crime, social exclusion and racial inequality. Community participation and the levering in of private sector and European resources were seen as particularly important elements of the regeneration process.

A proportion of the Single Regeneration Budget was 'top-sliced' to fund EP as well as the remaining urban development corporations and City Challenge initiatives until their closure (neither programme being seen as relevant to the new flexible locally driven style of regeneration). A further element was allocated, with some discretion given to the Regional Offices of Government, to fund Housing Action Trusts and the remaining Estate Action projects. The remainder was offered as a 'Challenge Fund' for which local agencies might bid. Within a budget that was declining in total through the late 1990s, the proportion available through the Challenge Fund increased from around 10 per cent of total SRB funds in 1995/6 to over 60 per cent in 1999/2000. Bidding for Challenge Fund support was competitive with central government determining the choice of projects it wished to support. Within the broad parameters of urban regeneration the government's priorities for the allocation of funds varied somewhat from year to year. For instance round four (1998/9) had a more explicit emphasis on tackling multiple deprivation than earlier rounds. Nevertheless the general principles remained constant. According to the government:

> The Challenge Fund is a catalyst for local regeneration. It complements or attracts other resources – private, public or voluntary. It helps improve local areas and enhance the quality of life of local people by tackling need, stimulating wealth creation and enhancing competitiveness. The activities it supports are intended to make a real and sustainable impact. It encourages partners to come together in a joint approach to meet local needs and priorities (DOE: SRB Bidding Guidance, 1993, p1).

Bids might be thematic or area based. Typical of the bids that won support were funding of renewal areas, refurbishment of council estates, town centre improvements, employment creation or training projects. Bids had to be made by a partnership of organisations with a stake in the proposal. This allowed local authorities to develop their facilitating role whilst prohibiting them from going it alone. There were annual rounds of bidding with proposals assessed against their ability to meet the government's regeneration aims, fit within local strategies, be well targeted, and achieve value for money. Thus central government retained a very large measure of control over the nature and location of public subsidy for regeneration projects. As with City Challenge this approach obliged local agencies to think clearly about their mission and priorities. Nevertheless the SRB Challenge Fund was criticised on the grounds that many projects that were too small to overcome structural economic problems, that power still lay with central government, that its decision making was inconsistent and opaque, and that the competitive bidding process resulted in much wasted effort and no funding for many worthy schemes (Couch, 2000, pp161-162).

Single Regeneration Budget in Liverpool

Liverpool City Council's response to the Single Regeneration Budget (SRB) was substantially informed by the fact that it already had Objective One status from the European Union. The allocation of funds to the region under Objective One was determined by a strategy outlined in what was known as the 'Single Programming Document', which stated, inter alia, that:

> A special feature of Merseyside is the very sharp degree of economic and social disparities within the region. These disparities are concentrated in well-defined localities. The effectiveness of the funds will be increased by targeting resources selectively on area of need, and areas of opportunity (Merseyside Single Programming Document, quoted in Gillespie C, 1998, p20).

To meet this need Liverpool City Council designated eleven area-based 'local partnerships' in the most deprived areas of the city. The boundaries of these areas were widely drawn and included 55 per cent of the whole population. This may have been spatial targeting but it was of the most imprecise kind. In each case the City Council was to be the lead partner with a designated senior officer providing liaison between the Council and the local partnership. The overall aim of the partnerships was to improve the economic well-being and quality of life of local residents.

The formation of the local partnerships was central to the SRB Challenge Fund bidding process in Liverpool. They brought together the key public, private and voluntary agencies and formulated the area-based strategies necessary to support successful SRB bids under the DOE's Challenge Fund regulations. Although round 1 of the annual SRB bidding process took place in 1994, before these partnerships had been established, linkages and plans were sufficiently well developed in two areas, Speke/Garston and Dingle, to allow successful bids to be made (Gillespie C, 1998, p23). Two contrasting programmes, Speke/Garston and the North Liverpool Partnership, illustrate the varying approaches that could be accommodated under the SRB Challenge Fund across the city.

Speke/Garston

The proposal was for a five-year programme of regeneration in the older residential areas adjoining the Garston Docks and the large, relatively self-contained, Speke housing estate. Garston developed in the 19th century as a working class housing area around the docks and nearby industrial areas. In the 1970s some council housing was located nearby. Despite being near the edge of the city, it had all the characteristics of a deprived inner urban area: poverty, unemployment, social problems and a degraded environment. Speke, containing more than 5,300 dwellings, was one of the largest council estates in the country, having been built mainly between the 1930s and 1950s. By 1994, with much reduced local employment and changing housing preferences, Speke housing estate showed evidence of severe social deprivation and needed substantial investment in order to refurbish the housing stock, infrastructure and local environment. The area also contained Liverpool Airport, the vacant 'old airfield' and a considerable amount of industrial development, including some of the largest manufacturing firms in the region, such as Ford, Dista, Glaxo Wellcome, and a Freightliner terminal at Garston Dock. Thus these two deprived residential areas adjoined one of the largest employment zones in the City, with some of the greatest potential for future investment and growth.

The Speke/Garston area had been put forward by the City Council in a bid for funding under round two of City Challenge. That bid had failed but when the SRB Challenge Fund was introduced only two years later the council and its partners were in a strong position to resubmit a somewhat modified bid for funding and were rewarded with £22 million of government money for a regeneration programme intended to last from 1995 until 2002. The strategy was:

> To provide a pathway for the people of Speke/Garston to achieve a 21st Century living and working environment by drawing on the abilities and commitment of local people, business and industry through a process of self regeneration (Liverpool City Council, 1995).

To achieve this, programmes were developed under three headings:

- *Action for local people:* including projects concerned with confidence building, training, welfare rights, equal opportunities, community health and crime prevention
- *Increasing the competitiveness of the private sector:* including support for indigenous businesses and attracting inward investment
- *Developing linkages:* including a labour market survey and skills audit, community transport, childcare, and a development trust

Figure 8.4   Liverpool John Lennon Airport
The new terminal building: part of the ongoing investment in the Speke/Garston area

Source: author

Complementing the SRB programme, the City Council and English Partnerships launched a joint venture, the Speke-Garston Development Company, to develop the economic potential of the area with the aim of attracting inward investment and expanding employment opportunities. The company was launched in 1996 with £14.5 million of European Union Objective One funding, with the intention of reclaiming the 146 hectares of derelict and vacant land in the area; providing up to 280,000m^2 of new commercial and industrial floorspace; and creating over 9,000 jobs. Its projects included:

- *The Estuary Commence Park* on the old airfield site (according to the company 'a superbly landscaped park-style business environment with striking water features suitable for office, distribution or manufacturing operations')
- *The Boulevard Industry Park* adjoining Ford/Jaguar's Halewood factory with a marketing strategy particularly aimed at automotive and pharmaceutical companies
- *The former Liverpool Airport terminal* which benefited from a £22 million scheme to convert the listed art deco building into a hotel and sports centre
- *The Match Factory* was a scheme promoted by developers Urban Splash to create a 'business village' in the former Bryant and May factory complex

Figure 8.5   **The Matchworks**
The former Bryant and May factory has been turned into an icon of economic recovery in the Garston area

Source: author

Regeneration activity in Speke/Garston was further widened in 1999 with a stock transfer of nearly 5,000 council dwellings to South Liverpool Housing (SLH), a new Registered Social Landlord (housing associated) charged with improving the condition and management of social housing in the area and funded with a grant of over £43 million from the government's Estate Renewal Challenge Fund. Rapid progress was made with housing improvements and in 2001 SLH, together with the Speke/Garston Development Company and the Speke/Garston Partnership won a best practice award from the British Urban Regeneration Association (BURA) for the overall regeneration of the area.

North Liverpool Partnership

In the second round SRB funding was secured for the North Liverpool Partnership (NLP): a six-year programme that claimed a strong emphasis on community-based economic develop. The Partnership included three districts within Liverpool's inner city: Breckfield, a predominantly residential area of private 'by-law' terraces and council housing. Everton, dominated by multi-storey council flats built to replace slums cleared in the post-war period; and Vauxhall, the mixed industrial and residential area adjoining the former north docks. In this run-down area of social deprivation and depressed environment a local partnership was established between the local authority, other public agencies, local businesses and the local community to bid for SRB support.

In the event £21.9 million was awarded for a six-year programme to create 'through effective partnership and the utilisation of the full potential of the whole community, a thriving area whose population enjoy good quality employment, education, health, housing and environment'. The tone and style of this vision statement was typical of many regeneration programmes but the emphasis on community involvement and social objectives was interesting. The grant was large by SRB standards but rather less than the £37.5 million awarded to five-year programmes for similar areas under the previous City Challenge initiative. According to the Partnership's strategy document the problems of the area were manifold. Educational attainment and aspirations were low, truancy and exclusions from school commonplace and youth and long-term unemployment endemic. There was seen to be a need for very personalised forms of basic skills training. Small local firms needed better access to sources of capital, business contracts and marketing in order to expand. Much of the housing stock was of poor quality. There were 15 former council owned tower blocks (many in a poor state of repair), that had been taken over by the

Liverpool Housing Action Trust. Fear of crime was a major issue. Everton Park, a large open area created by the City Council on former housing land, had been developed over the previous two decades but had always lacked the investment needed to make it a useful amenity. Nevertheless, with its stunning views over the Mersey and North Wales, it had the potential to be a major recreational asset.

NLP proposed to translate the strategic vision into action through a series of programmes referred to as 'routes':

- *Routes for People* developed policies and projects for education and training, employment, housing renovation and management, the delivery of healthcare and public transport
- *Routes for Business* targeted the expansion of indigenous local businesses, development of new firms and inward investment
- *Routes to Partnership* sought to improve relationships and co-ordination between the multiplicity of agencies and stakeholders

NLP was managed by a steering group comprising representatives of local residents (6 members), local businesses (6 members), the public sector (6 members), 4 ward councillors and a Chair appointed by the City Council. The work of the Partnership was carried out by an Executive Team including a manager, community co-ordinator, two community development workers and a number of other officers. Through the six-year strategy it was anticipated that SRB funding would be complemented by other public sector funds including £36 million from LHAT, £16 million from the Housing Corporation, £12 million in European funding, £2.9 million from the City Council, and £2.5 million from English Partnerships (NWDA). It was hoped that this investment would then lever in more than £43 million of private funding. Thus the total investment expected in the area over the six years was in the order of £138 million.

This strategy was intended to be a comprehensive and holistic attack on the problems of the area and as such it represented a continuation and development of the approach to regeneration that emerged in the thinking behind renewal areas (1989) and City Challenge (1991). The co-ordination of spending programmes by public sector agencies was a key part of the strategy. This reflected the lessons learned from the government sponsored Task Forces and City Action Teams in the eighties and the approach signalled by government in establishing their Regional Offices of Government. The levering in of private investment had been seen as an important measure of urban regeneration achievement since the early days of the urban development corporations.

In the third round of the SRB Challenge Fund the DOE supported further regeneration programmes within the city including the Netherley Valley Partnership, a six-year project on two large deprived outer council estates and aimed at strengthening the local economic base; and the Liverpool East Area Partnership, a five-year project that was to concentrate on refurbishing and developing four neighbourhood (shopping) centres improving local access to training and employment. By the fourth round of bidding the City Council had decided to avoid the growing local competition for resources by putting forward a thematic bid to improve primary education in the partnership areas. At the same time, although not directly funded through the SRN Challenge Fund and established on a different basis to other partnership areas, the Ropewalks Partnership provided another example of a co-ordinated response to a regeneration problem with the City Council adopting a facilitating role.

Rope Walks

The Duke Street/Bold Street area adjoins the city centre and the south docks. Until recently the area contained a mixture of port related uses, warehousing and small scale industrial activity, and fringe city centre uses such as secondary shops, discount stores, offices and workshops. By the 1980s the area had become very run down. With the closure of the south docks many of the warehouses and workshops became derelict. Much of the limited housing stock was abandoned. In Bold Street itself, pedestrianised at the city centre end, many shops stood empty and upper floor vacancy was commonplace. Even the roads and pavements were in a perilous state. Yet the area clearly had potential. It was compact and well defined. There was a unity and human scale about its urban form. The streets were narrow and the buildings generally no more than 3 or 4 storeys in height. There were some Georgian houses (though few in their original use) and some fine Victorian warehouses.

By 1988 the City Council had recognised the architectural and historic merits of the area and designated the Duke Street/Bold Street Conservation Area. Although covering around 1 km^2 the area was estimated to have less than 1,000 residents at that time. Shortly afterwards the Council sold many of its properties in the area to Charterhouse Estates Ltd, who promised to create, in partnership with the City Council, a 'creative industries quarter'. This was to be a mixed-use area of studios and workshops, galleries and performance spaces, bars and restaurants, specialist shops and above all – housing. Some of the redundant buildings would be renovated whilst others were to be cleared for redevelopment. The public realm was to be improved and new streets created. Most of the streets ran in a parallel east-west direction and there were few north-south connections making it difficult to penetrate the district. The idea was to make it easier for pedestrians to move around the area, i.e. to improve what urban designers refer to as permeability. As luck would have it the property recession of 1991 claimed Charterhouse Estates Ltd as a victim and the firm went into receivership. The vision had stalled.

Despite this setback the City Council continued to promote the 'creative quarter' but little happened until the mid-nineties when the Council re-paved and redesigned the public realm of Bold Street and Urban Splash (then a little known local firm) redeveloped Concert Square, between Fleet Street and Wood Street, with a new public space, two bars, workspace and living accommodation above. This proved to be a groundbreaking development. Firstly this was a new kind of firm, combining property development and design skills and investing in a high risk and difficult project in a marginal location surrounded by dereliction. Being the first significant developer in the area required an act of faith and a belief that the concept of a 'creative quarter' was more than just hype. Secondly, this was a mixed-use project combining bars, workspace and apartments in the same building: a concept that was novel and virtually untested in British cities at the time.

At about the same time the City Council, working with English Partnerships, commissioned preparation of a Duke Street/Bold Street regeneration strategy. The final strategy, known as the Integrated Action Plan (IAP), was published in 1997 after an extensive community participation process. To co-ordinate the implementation of such a complex area a new organisation was established: the Rope Walks Partnership. The name derived from the area's historic association with rope making and chandlery, although it would not have been recognised by any local people at the time. The Rope Walks Partnership Board comprised members drawn from the City Council, English Partnerships, the voluntary sector and local community. This Board was supported by an executive team of seven professional staff (Couch and Dennemann, 2000, p142).

The IAP was based on the ideas that had first been proposed at the time of the Charterhouse initiative eight years earlier. The strategy built upon the existing business structure with an emphasis on the development of creative industries, culture and the night-time economy. Mixed uses were strongly encouraged. Improvements to the public realm were also proposed. Streetscape was to be improved and buildings refurbished to create a safe and sustainable environment in which people would choose to live and work. A number of specific development projects were identified including Concert Square, the 'Tea Factory' and FACT centre, Wolstenholme Square, Chinatown, Dukes Terrace and Henry Street.

Although the partnership was only scheduled to last for five years, there has been a remarkable transformation of the area that now looks as if it has the real potential to develop into an exciting and thriving mixed-use urban quarter. One of the key flagship developments has been the FACT Centre (Foundation for Art and Creative Technology). This was the redevelopment of a site in Wood Street for a film and arts centre, supported by lottery funding. It opened in the winter of 2002/3 with the hope that it would be catalytic cultural attraction drawing many visitors into the area.

Adjoining FACT the new Rope Walks Square links into the east end of Bold Street. Further west along Wood Street another new square provided a link southwards to Seel Street. Throughout the district by 2002 there were new retail outlets, bars, restaurants, night spots, studios, galleries, workspaces and above all: housing. By the end of 2001 more than 400 apartments had been completed or were in the process of development.

In the five years that followed their initial foray into the area, Urban Splash grew to become a major player in central area regeneration projects in Liverpool, Manchester and beyond. The city centre has become a well-established location for private housing investment. Mixed-use projects are no longer feared by developers as they once were. And the redevelopment of the Rope Walks seems to have reached sufficient critical mass as to become economically sustainable into the future. The area has become in many ways a model of successful partnership working in urban regeneration. (See Couch and Dennemann, 2000 for a more detailed account of the redevelopment of the area.)

Figure 8.6   **FACT**

A new arts centre and flagship project for the Rope Walks. A series of new pedestrian ways (including Rope Walks Square – foreground) increase permeability through the area

Source: author

Figure 8.7   Rope Walks housing
A former warehouse converted into private apartments in Fleet
Street in the heart of the Rope Walks

Source: author

Regional and Strategic Frameworks

European funding

In 1993 Merseyside was designated an Objective One region for EU funding on the basis that it was an under-performing region with a gross domestic product (the amount of economic activity) per head of population being less than 75 per cent of the EU average. A programme for economic regeneration was contained within a 'single programming document' drawn up under the auspices of the Government Office for Merseyside and was known as Merseyside 2000.

The vision of this programme was to establish Merseyside as a prosperous European City Region with a diverse economic base (European Commission, 1994, p25). Five key drivers were identified as:

- action for industry (inward investment and key corporate business development)
- action for industry (promoting indigenous enterprise and local business development)
- action for industry (knowledge-based industries and advanced technology development)
- action for industry (developing the cultural, media and leisure industries)
- action for people (pathways to integration, a better training system, community development and a better quality of life).

The need to develop new technology industries was recognised. It was also significant that the cultural, media and leisure sectors were identified for their importance to the regional economy, including the concentration of museums and galleries in central Liverpool, and the rapidly growing tourism industry. A special feature of the area identified by the programme was the sharp degree of social and economic disparities in the region. This was particularly relevant to the fifth driver, action for people, where there was to be a spatial focus on areas worst affected by long-term unemployment and low incomes. A number of parts of inner Liverpool were identified as such areas. In these areas the priorities were community development and involvement, career development, equal opportunities, improving access to jobs and training for facing exclusion from the labour market, improved and more flexible education, better training and employment services, improved public transport, environmental improvements and derelict land reclamation (European Commission, 1994, Ch5).

Implementation began slowly and it was only towards the end of the funding period that changes in the area began to be visible. Towards the end of the decade the national economy was improving an unemployment falling, not least in Liverpool. There was an upturn in housing investment, including the emerging phenomenon on housing development in the city centre itself. Much of the European funding was spend jointly funding schemes that may have had their origins in earlier regeneration initiatives. By the millennium there was a perceptible increase in investments in the city centre, the pathways (partnership) areas and the economic regeneration areas such as the eastern Corridor, Waterfront, Gillmoss/Fazakerley/Aintree and Speke/Garston.

Nevertheless, the local economy continued to under-perform relative to the EU average and Merseyside was designated to receive a second tranch of Objective One funding between 2000 and 2006. The analysis of local economic circumstances remained similar to that which had been presented six years earlier. The vision, however, had changed to one of achieving 'a world class city-region that attracts people to live work, invest and visit' (European Commission, 2000). This was a significant change. It emphasised the importance of Liverpool and especially Liverpool city centre, in the economy of the region. In so doing it also recognised the importance of service industries, including consumption activities such as shopping, recreation, culture and tourism, to the economy of modern city regions.

The priorities had also subtly changed. The first priority was developing business where the aim was 'the creation of a competitive, knowledge driver economy, focussed on entrepreneurship, business competitiveness, and world-class management skills'. Developing people's employability and skills, and 'developing pathways communities' remained important objectives but there was a new emphasis on 'developing locations' and the creation of a competitive environment in order to attract inward investment. The development of information and communication technology (ICT) had become a key cross-cutting theme along with equal opportunities, social inclusion and environmental sustainability. This was a much more advanced programme that which has been approved in 1994, reflecting the modern role of city-regions and more in tune with contemporary urban policies across Europe.

Regional strategies

To co-ordinate planning in the region the North West Regional Association in 1994 published 'Greener Growth' setting out its advice to the government on a broad framework for development in the region over the next two decades (North West Regional Association, 1994). The document was intended to form the basis for regional planning guidance. The topics covered included the regional environment, economy, housing and transport. Despite purporting to have a central theme of 'greener' or more environmentally sustainable development and growth, the reality of the plan is that it is more concerned with economic growth than environmental protection. The Association lacked the teeth to produce a strongly focussed plan that could strongly push forward the environmental agenda. Despite recognising the importance of urban regeneration the plan allows considerable flexibility for local authorities to permit peripheral growth. Only limited support is given to public transport improvements whereas there are a number of circumstances in which trunk road schemes would be prioritised.

Responsibility for regional planning guidance passed to the new regional assembly who, in 2000, published new draft regional planning guidance (North West Regional Assembly, 2000). This guidance established a spatial development framework for the region that aimed to achieve:

- sustainable patterns of growth and change across the region;
- concentrating growth and change in regional centres and towns.

There were different principles to govern the development of each part of the region. Core strategies aimed at economy in the use of land, enhancing existing capital, and quality in new development. Each of these, especially the first two, had significant potential benefits for Liverpool: reinforcing the commitment to restrict peripheral growth and encouraging urban regeneration. Specific policies affecting Liverpool included:

- concentration of development and urban renaissance resources on the conurbations of Greater Manchester and Merseyside and, inter alia, on Liverpool's inner core and city centre;
- protection of Liverpool city centre as a regional shopping centre;
- protection and conservation of the built heritage of Liverpool's commercial centre and waterfront;

- exploitation of the regeneration potential of Victorian and Edwardian commercial developments in Liverpool city centre;
- identification of the Speke/Garston area as a regional investment site;
- recognition and support for the port of Liverpool as the North West's key international seaport;
- increasing the re-use suitable vacant urban land and buildings for housing (Liverpool was expected to provide for 26,100 new dwellings between 1996 and 2016 (9 per cent of the regional total);
- development and improvement of urban public transport infrastructure and services.

None of these ambitions were new to the city but it was useful to have the backing and support of the region as a way of strengthening and building confidence these policies.

At the same time as regional planning guidance was being formulated the North West Development Agency was preparing its own strategy. The ideas it contained were by now familiar: promoting regional competitiveness; encouraging sustainable communities and social inclusion; emphasising investment in ICT; improving the workforce; working through partnership and improving efficiency in the use of resources (North West Development Agency, 2000). The two strands of regional policy development were brought together in a report for the DETR (DTZ Pieda, 2000). This sought to establish the spatial implications of the North West regional strategy and raised some serious questions that highlighted the tension market forces and the idea of sustainable development.

> The desired locations for these (growth) sectors are high quality campus/business parks. The main mode of travel to work is...by car and so potentially not environmentally sustainable...There is relatively little correspondence between desired locations and the strategy's areas of priority regeneration need...Generally, there is a fundamental mismatch between the desired direction of transport policy (less road borne movement of goods and employees) which RPG focuses on, and the current locational preferences of many sectors (DTZ Pieda, 2000, p3).

Thus whilst the regional strategy appeared to be supportive of the regeneration of Liverpool, there were serious questions about the ability of government agencies to deliver a strategy that diverged so much from market requirements. Real tensions existed between the desire for economic growth in the Liverpool city-region and the need for environmentally sustainable development. The former was perceived to have greater short-term political and social benefits whilst the latter was of fundamental importance in the longer-term.

New Labour and Urban Regeneration

Whereas it had been the strategy of John Major's government to streamline and simplify regeneration funding and processes, the first years of the new Labour government seemed to take policy in the opposite direction. A Social Exclusion Unit and Policy Action Teams were established in central government. Urban policy was distinguished from neighbourhood renewal within a department that evolved from the former Department of the Environment. Initially this new Department of Transport, Local Government and the Regions (later styled the Office of the Deputy Prime Minister) included transport but later this was hived off back to its own separate department. Most of the functions of English Partnerships were passed to the new Regional Development Agencies (RDA). However English Partnerships continued to exist, mainly to undertake and advise on brownfield development, to manage and develop surplus government owned land and to run a series of special programmes including the millennium villages programme and the urban regeneration companies (including Liverpool Vision). The RDAs themselves were responsible to the Department of Trade and Industry. Thus economic regeneration was to become somewhat detached from neighbourhood renewal and urban policy, at least at central government level.

Initially the government set up the New Deal for Communities but then came up with a more comprehensive National Strategy for Neighbourhood Renewal and a Neighbourhood Renewal Fund to replace the SRB Challenge Fund. The strategy was to reduce unemployment, low skills, crime, poor health, improve housing and physical environments in the poorest neighbourhoods. It was also hoped to narrow the gap between these neighbourhoods and the rest of the country. One new feature of policy was that much more emphasis was now to be put on bending mainstream policies towards deprived areas than had previously been the case. It was argued that by better targeting of mainstream programmes more resources could be channelled into these areas for a more sustained period than had previously been achieved through special programmes of additional funding. This approach was reminiscent of the short-lived approach to urban policy adopted by the Labour government after 1978.

The government identified 88 of the most deprived local authorities (including Liverpool) to be beneficiaries of Neighbourhood Renewal

Funding subject to a local Renewal Strategy to be drawn up by a new Local Strategic Partnership (representing the local authority together with other statutory, voluntary, business and community organisations).

New Deal for Communities: Kensington

One of the first outcomes of the incoming government's review of urban regeneration policy was a new programme to tackle multiple deprivation in the most deprived neighbourhoods in the country in an intensive and co-ordinated manner. This was known as the New Deal for Communities (NDC) and its aim was to bridge the gap in performance between these neighbourhoods and the rest of the country against a series of indicators.

The programme built on past experience of neighbourhood regeneration and had many of the characteristics of the recent renewal areas and programmes funded under the SRB Challenge Fund. It was to be holistic and partnership based but in a change with recent practice there would be no bidding process: money would be allocated by central government.

The Kensington district of Liverpool was announced as one of the first round of NDC areas in 1998. It included 4,200 households in Kensington, Fairfield, Edge Hill and Wavertree. Key features of the regeneration process included community involvement, partnership and 'joined-up thinking' (i.e. co-ordinated action between departments and agencies), an evidence-based approach to policy formulation, and a long-term commitment to deliver real change in the area. To support this programme the government committed £61 million, significantly more than had been available to individual SRB funded projects or even in the City Challenge areas earlier in the decade.

The regeneration programme in Kensington included housing improvements, the clearance and replacement of outworn or abandoned housing, and environmental enhancements. One novel project was that of 'Wired-up Communities' intended to provide 2,000 people with recycled computers, internet access and ICT training.

The district is bisected by Edge Lane, an important radial artery linking the city centre with the national motorway network (M62). An important issue for the NDC was to therefore to consider the future of this road taking into account, on the one hand, the desire to provide a prestigious and efficient road link from the region, through Kensington, into the city centre, and on the other hand, the need to minimise traffic intrusion into the area and to discourage road borne traffic movements. It is too early to comment on the outcomes of the Kensington NDC and its impact on the area but the scale of the programme is indicative of a serious commitment to change.

Liverpool Vision

One of the many proposals to emerge from the report of the Urban Task Force (UTF), *Towards an Urban Renaissance* (Urban Task Force, 1999) was a call for the establishment of 'urban regeneration companies'. In response the government initially established three such companies: Liverpool Vision in June 1999, New East Manchester in October 1999 and Sheffield One in February 2000. Although these companies were not precisely based on the UTF model they did represent a new development in urban regeneration.

Liverpool Vision was to be a limited company charged with preparing and implementing proposals for the regeneration of Liverpool City Centre and the major transport corridors into the centre. Initially it was to be funded by English Partnerships, the North-West Development Agency and Liverpool City Council. Board members included representatives from Wimpey, Tesco, Littlewoods, Liverpool Stores Committee and Maritime Housing Association. The core staff was small with much of the work sub-contracted to consultants.

> Most of the first 12 months was spent drawing up the strategic regeneration framework, mapping out how the city centre should be developed over the next ten years. Its strategic goals are loaded wit buzz-phrases like 'a high-quality safe urban environment', a 21st century economy', 'a world-class tourist destination', 'a premier national shopping destination', and 'inclusive communities' (Brauner , 2000, p21).

The idea of urban regeneration companies had been broadly welcomed in a report for the Department of the Environment, Transport and the Regions (Robson B and Parkinson M, 2000). However, the concept raised a series of questions, at least insofar as Liverpool Vision was concerned. One issue was that of the relationship between the company and the local authority. For a number of years Liverpool City Council had been painstakingly preparing its UDP, which included a detailed analysis and proposals for the city centre. It could be argued that whilst these proposals represented a balanced view of the future role and planning of the city centre in relation to the needs of the community as a whole, the proposals from Liverpool Vision, which did not precisely coincide with the UDP, were derived from a narrower perspective. Furthermore, whilst Liverpool

Vision included local authority representatives on its board, the organisation was not directly subject to local democratic control. This was very different from the original UTF proposal:

> an Urban Regeneration Company should be capable of acting swiftly, as a single purpose delivery body to lead and co-ordinate the regeneration of neighbourhoods in accordance with the objectives of a wider local strategy which has been developed by the local authority and its partners (UTF, 1999, p147).

The actual Liverpool Vision 'master plan framework' was prepared in 1999 by consultants Skidmore, Owens and Merril. Its primary aim was to enhance the environmental quality of the city centre and to bring about investment in the area. The city centre was divided into six zones that provided a structure for intervention. These included 'Castle Street Live-Work District', 'Cultural Quarter/Lime Street Station' and 'Pier Head'. A set of design principles was established to guide future developments. These included connecting the city centre with adjacent communities and extending the pedestrian priority zone and improving the sense of 'arrival' by upgrading and developing the city centre gateways. Other ambitions included streetscape improvements, the redesign and creation of new squares, mixed-use developments, which were claimed to be the key to promoting the concept of the 24-hour city that in turn was said to ensure streets would be active and safe. Also important was the protection and restoration of the city centre's impressive heritage environments and buildings. Despite the political hype surrounding the plan, it did not offer a great deal that had not already been proposed by the City Council in its own planning documents. What it did offer, as the UDCs and other agencies had done before, was the political focus, the access to funding and the machinery to stimulate and promote development.

Conclusions

The new decade had seen a move away from property-led regeneration towards something more holistic. In many ways this was a return to the 'total' approach that had been recommended by various reports and studies back in the nineteen seventies. Local authorities were also brought back into the picture: not so much as deliverers of service, which had been their traditional role, but as facilitators and co-ordinators. In this new era, when so much service delivery was in the hands of different organisations (government departments, regeneration agencies, housing associations, privatised utility and transport companies) the need for leadership and co-ordination in urban regeneration was becoming a major issue. Local authorities had the capacity to undertake this emerging role, even if their funding and their emasculated powers only allowed them a modest direct role in implementation.

From the renewal areas onwards the government showed a new commitment to local partnership, community and stakeholder participation in urban regeneration. In City Challenge and SRF Challenge Fund programmes local authorities were expected to consult with local communities and to act in partnership with other agencies. LHAT was particularly impressive in its support for capacity building amongst tenants and engaging its tenants in policy-making and implementation. However, there were complaints that some aspects of urban regeneration, notably some English Partnerships programmes, were still heavily top-down in style. This ambivalent attitude continued with the new Labour government giving strong encouragement to community participation in local authority led partnerships programmes whilst supporting urban regeneration companies with little provision for local accountability.

Though the decade there was a shift in emphasis in economic regeneration recognising the increasingly international and global nature of investment and the need for cities and city-regions to be competitive. In this context there was a growing acknowledgement of the importance of Liverpool city centre to the regional economy and the need to bring its environment, amenities and facilities up to an internationally competitive standard. Building on local economic strengths was also seen to be important with agencies offering co-ordinated action in support of the ICT and biotechnology industries as well as the culture, recreation and tourism sectors.

However, during the decade there emerged a growing tension between the aims of economic growth and environmentally sustainable development. With the gathering momentum of the environmental agenda and the growing confidence of the environmental movement after 1990, economic growth objectives and strategies were more frequently challenged. Increasingly, albeit too slowly for many on the environmental side, regeneration plans and programmes began to take on board environmental considerations. By the end of the decade, with the support of the central government, the North West regional development strategy, the Merseyside Objective One Single Programming Document, the Liverpool Unitary Development Plan and various local area initiatives all had sustainable development and social inclusion as key aims.

9 Planned and Sustainable Regeneration?

The Evolution of Policy

Since the 1960s, Liverpool's role in the national and international economy has diminished with consequences for social and environmental conditions. Economic development, social inclusion and environmental improvements have become key challenges facing policy makers. The IPPS of 1965 was an important urban planning and corporate policy document for the City Council. Taking a comprehensive view of the city as an interconnected urban system, the plan reflected and was facilitated by the fact that the Council itself still retained responsibility for the great majority of local public services in the area. The main aim of planning at this time was to modernise the physical form and structure of the city, especially provision of a radical new infrastructure for urban transportation.

By the end of the decade urban deprivation had become an issue of concern. A series of studies and experiments gradually led to the conclusion that the main underlying cause of these problems lay in the impact on local areas of external decisions and structural economic changes, beyond the control of local people. It was clear that economic regeneration was a prerequisite to tackling urban deprivation and that regeneration policy had to be holistic and deal with all the causes of urban deprivation at the same time. The necessary co-ordination between the various agencies and levels of government could be achieved through working in partnership. The 1978 Inner Urban Areas Act and associated policy changes reflected this approach but were not politically robust and were in large measure abandoned by the incoming Conservative government.

During the seventies the statutory system of development plans became undermined by a series of events, not least the reorganisation of local government in 1974, the effect which was to fragment responsibility for different elements of urban policy between different levels and agencies of government. Nevertheless, two important strategic plans did emerge during this time. The Strategic Plan for the North West sought to concentrate regional development in the Mersey Belt between Liverpool and Manchester and was significant for its recognition of the emerging environmental agenda. The Merseyside Structure Plan established the twin policies of investment in the inner areas and restrictions on peripheral growth as the basis for urban regeneration in Liverpool.

Coming to power in 1979 the Thatcher Government adopted a strongly market-orientated and property-led approach to urban regeneration, quickly establishing the Merseyside Development Corporation and Speke Enterprise Zone while abandoning the idea of partnership with the local authorities. This might have been acceptable had it not been accompanied by substantial cuts in rate support grant that forced many urban local authorities to curtail regeneration initiatives and reduce spending on mainstream programmes. This approach to policy brought conflict. There were inner city riots and confrontations between central and local government. Reforms were brought in by central government to prevent subversion of their agenda. Only at the end of the decade was there a return to a more conciliatory and inclusive approach to urban policy making. Nevertheless, regeneration in this period was not without successes. The Merseyside Task Force brought a useful element of co-ordination to government decision-making. The development of Wavertree Technology Park brought significant local economic benefits. The validity of bottom-up approaches to housing development was demonstrated through the Eldonians and others in the 'Vauxhall Villages'. Merseyside Development Corporation achieved much of what it set out to do in redeveloping the former docklands and their surrounding areas. However, these ad-hoc and opportunistic regeneration initiatives tended to have little regard to wider plans for the development of the city. Within the City Council plan making had been frustrated by slow progress on the structure plan and internal political difficulties. By the 1980s economic development, urban regeneration and environmental improvement were the dominant planning concerns in Liverpool. Transport planning had slipped down the agenda.

By the time the Unitary Development Plan was finally published, responsibility for so much policy making had passed to other agencies that it was a much less important policy document than its predecessors. Nevertheless, it did attempt to widen and modernise the planning agenda. The UDP showed an understanding of sustainable development that surpassed that of previous plans and its development control policies were more sophisticated than any that had preceded.

The 1990s saw a move away from property-led regeneration towards a more holistic approach. The City Council was given renewed responsibilities, not so much as a deliverer of services but as a facilitator and co-ordinator. The government showed a new commitment to local partnership, community and stakeholder participation in urban regeneration. In City Challenge and SRB Challenge Fund programmes the local authority was expected to consult widely and to work in partnership with other agencies. However, some regeneration programmes, notably those of English Partnerships/NWDA remained top-down in style. This ambivalence towards participation continued with the post-1997 Labour government giving strong encouragement to community participation in local authority led partnerships and programmes whilst supporting urban regeneration companies with little provision for local accountability.

Sustainable development emerged as one of the key aims of contemporary urban policy and there was a growing tension between the needs of economic growth and the environment. Economic development strategies were increasingly challenged and gradually urban plans and regeneration programmes began to take on board the environmental agenda. Also through the decade there was a shift in emphasis in economic regeneration recognising the increasingly international and global nature of competition for inward investment and the importance of cities as centres of service employment and consumption. In this context there was a growing acknowledgement of the importance of Liverpool city centre to the regional economy and the need to bring its environment, amenities and facilities up to an internationally competitive standard. By the millennium the City Council, working with Liverpool Vision and others was working towards such a goal.

The Regeneration of Liverpool

Although there has been continuing debate about urban regeneration policies implemented in Liverpool over the past forty years and the city remains near the bottom of deprivation league tables, the remarkable feature of urban policy in Liverpool is perhaps not how much it has failed, but how much it has succeeded. Yes, the city has problems, but they are not vastly different from any other conurbation. Despite losing more than 40 per cent of its population and 17 per cent of its households since 1961, a recent analysis showed vacant dwellings to be only 5-6 per cent of the total stock: a proportion not dissimilar from many other large cities (Liverpool City Council, 1998, p8). Through a combination of the effects of declining household size and the demolition of unpopular housing, the overall relationship between housing demand and supply has been kept in a reasonably approximate balance. This is not a ghost town or a desert. The city has adapted to change, principally by reducing densities. The only exceptions have been in neighbourhoods with specific problems, either containing unpopular dwelling types or becoming stigmatised through crime or other problems.

Furthermore, there has been a major improvement in living conditions over the period. The proportion of households in Liverpool without exclusive use of all amenities (bath, inside WC, etc.) has fallen from around 30 per cent in 1971 to around only 1 per cent today. Admittedly the state of repair of many dwellings leaves much to be desired but this real progress should not be overlooked. Even in the nineteen sixties most dwellings in the inner areas relied on coal fires. More than ten thousand domestic chimneys would pour out smoke on a winter's day. Today, even on a cold and bright January day scarcely a single smoking chimney can be seen: the air is clear.

Between the beginning of the 1960s and the millennium the number of jobs in the city fell by about a half. Although unemployment rose dramatically and with serious social consequences, in the 1970s and 1980s, there was a fairly consistent trend for the unemployment rate in the city to remain at around twice the national average. This is, of course, unacceptably high and a challenge to policy makers, but the point is that it could have been much worse. Through a combination of the effects of out-migration of people seeking work and the job-creating effects of economic regeneration policies, labour supply and demand have shown a long-run tendency to remain in a reasonably stable relationship one to the other. Another effect of industrial closure and technological change in the 1970s and 1980s was the creation of great tracts of vacant and derelict land. At the peak of the problem in the late 1970s around 750 hectares of vacant and derelict land were recorded in the city (Merseyside County Council, 1978, p146). Notwithstanding a continuing process of abandonment creating dereliction, urban regeneration policies have reduced the amount of vacant and derelict land to a little over half that figure today (Liverpool City Council, 2002).

Thus any debate about the effectiveness and efficiency of urban regeneration policies should take place within the context of an understanding that the urban system has managed to adjust to maintain an approximate long-run balance between changes in the demand and supply of housing, employment and land. This is really quite impressive and yet easily forgotten. The fact is that things could have been much worse in Liverpool over the last thirty years than they actually have been. Clearly much of this adjustment has occurred through the workings of the market but urban policy interventions have also played an important part:

particularly in relation to the re-use and redevelopment of land and buildings.

Nevertheless beyond these headlines the detailed picture of regeneration is a little more complex and the level of success rather more mixed. Perhaps one of the key measures of success in regeneration from a welfare perspective is that of convergence. Many indicators of deprivation and social exclusion place great store on the measurement of differences between the situation in any given area and local or national norms. Hence, for example, Merseyside receives Objective One funding because its Gross Domestic Product (GDP) per capita is below 75 per cent of the EU average. Therefore if we can measure the extent to which the conditions in a particular area have improved relative to the average, we can say something about the degree to which regeneration has been successful. The issue can be considered at two spatial scales: a comparison between the city of Liverpool with other cities and with national trends; and a comparison between different areas within Liverpool with city-wide norms.

The first question to be considered then is how has Liverpool fared in comparison with the rest of the country? If success in regeneration is to be measured in terms of convergence, it will not be measured by the absolute amount of change that has occurred in Liverpool but by the way Liverpool has changed relative to other areas. There are few indicators that are representative of the regeneration process and available on a comparative basis for different cities and the country as a whole over a long enough time scale to be of value in this exercise. Bearing these constraints in mind one of the key indicators of regeneration is population change. Population decline is an outcome of the interaction of many adverse circumstances including economic decline, social deprivation and environmental degradation. Population growth occurs in regions and sub-regions experiencing economic growth and within those areas, population will be attracted to districts where social and environmental conditions are good. Figure 9.1 shows the rate of population change in selected English cities since 1971.

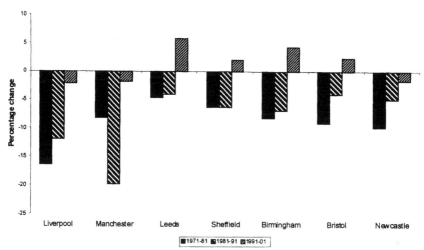

Figure 9.1   Population change in selected English cities

Source: Census of Population

Some caution needs to be exercised when reading this figure. The data is presented for local authority areas. The nature of these areas varies significantly between cities. For instance, in Leeds and Sheffield the boundaries are drawn very loosely and include large areas of suburb and countryside, whereas in both Liverpool and Manchester the urbanised area extends well beyond the local authority boundary. The effect of this difference is that in some cities, such as Leeds and Sheffield, much of the sub-urbanisation and decentralisation process of recent decades has been occurring within the city boundary, whereas in Liverpool and Manchester much of this process has been taking place beyond the city boundary.

Nevertheless, in the period 1971-81 Liverpool suffered more population loss than any other comparable English city. In the period from 1981-91 it continued to suffer more population loss than any other British city except Manchester, and in the nineteen nineties when most cities were beginning to regain population, Liverpool continued to perform less well than its counterparts.

To the extent that population gain is evidence of regeneration it can be supposed that all these cities have experienced some regenerative effect during the last decade but that its extent has varied between places. Clearly regional economic circumstances have played a significant part in local success as the highest rates of population growth have been in Birmingham and Bristol in the economically prosperous West Midlands and South West (where GDP per capita is 94 per cent and 100 per cent of the EU average respectively). Leeds has also performed well. Sheffield lies in South

Yorkshire, an area that recently qualified for EU Objective One status. However, its loosely drawn boundary may have inflated the real extent to which the urban core has been re-populated. Newcastle has also performed relatively poorly although improving. Despite being a regional capital, the sub-regional economy is weak and the local authority boundaries are tightly drawn. Manchester, despite being a regional capital with a relatively strong economy continues to be affected by decentralisation and performs relatively badly. Liverpool is the worst performer. Whilst there has been some re-occupation of the city centre in recent years, the city as a whole continues to lose population. Regeneration has been limited by the continuing weaknesses in the Merseyside economy when compared with national trends.

At the scale of the conurbation a comparison of population trends in Liverpool with those of the whole Merseyside county area provides an indication of the extent to which suburbanisation and decentralisation continue to be problematic. A reversal, or at least a slowing in the long-term trend towards decentralisation would provide an indication of the success of regeneration in Liverpool.

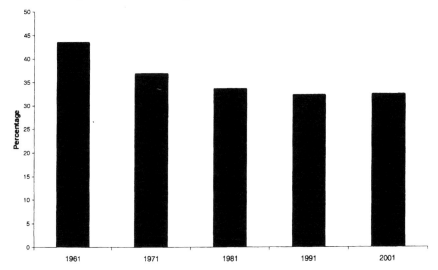

**Figure 9.2   The proportion of Merseyside population living in
                  Liverpool**

Source: Census of Population

Figure 9.2 indicates the proportion of the Merseyside County population living in Liverpool. Whilst decentralisation continued throughout most of the study period, well into the 1990s, the rate of decentralisation was slowing. This pattern coincides with changes in policy. During the 1960s slum clearance and the building of overspill council estates, new towns, expanded towns and private speculative suburban housebuilding was in full swing. By the 1970s clearance was slowing and few new overspill schemes were being started. By the 1980s the clearance of the older private housing stock had effectively come to a halt although there was some removal of obsolete council stock through Estate Action and later, by LHAT. By the mid-1980s private speculative suburban housebuilding was being severely restricted by policy, and there was a growing amount of additional housing being provided within the city through both new build and conversion schemes. However, at the same time average household size had been falling and because the inner areas rather than the suburbs tended to attract more students and smaller, younger households, average household size in Liverpool consistently remained below County average. This exacerbated the difficulties facing the inner areas in retaining population. Thus for Liverpool to show in the 1990s some slight reversal of the previous trend towards decentralisation, is evidence that Liverpool may be beginning to regain its position within the Merseyside conurbation.

Another important indication that regeneration has taken place is the degree of convergence between different parts of the city over time. The more regeneration has been successful and well targeted, the more formerly deprived areas should converge toward the city norms. There are a limited number of indicators that are useful in this regard and consistently available for small areas over a sufficiently long period of time to be meaningful. As has already been suggested, population change within different parts of the city is another important indicator of the success of regeneration. Figure 9.3 shows data for selected wards including those that suffered the highest rate of population loss between 1971 and 1981 and compares their performance through the subsequent decades.

In most of these wards the decrease in population between 1971 and 1981 was brought about by a combination of natural change, net out-migration and falling household size, exacerbated by the continuation of large scale slum clearance. One exception to this pattern was Netherley where, despite consisting of mainly new council accommodation, the unpopularity of the estate was such that some loss of population still occurred during the decade. By the 1980s, despite a virtual end to clearance, population was haemorrhaging from both the inner area council estates that dominated housing provision in Vauxhall and Everton and the remote peripheral council estates, of which Netherley was the most extreme example. However, by the 1990s each of these wards shows some evidence of change. All perform better than in previous decades with reduced rates of population loss indicating that regeneration policies have brought the problem under control. In one ward, Abercromby, the losses of previous decades have turned to population growth. This is no surprise. Abercromby, which lies immediately to the south-east of the city centre, includes the northern part of the former south docks, the Ropewalks area, the cathedrals, universities and Canning conservation area. The area benefited from MDC, City Challenge, EP and other regeneration funds. During the period 1981-91 it has been calculated that the ward received over £140 million investment, the highest of any inner city ward by a considerable margin (Department of the Environment, 1994, p269). Furthermore, the area continued to benefit from substantial public funds and renewed private sector interest in the subsequent decade. Evidently this concentration of regeneration investment over a sustained period has brought results.

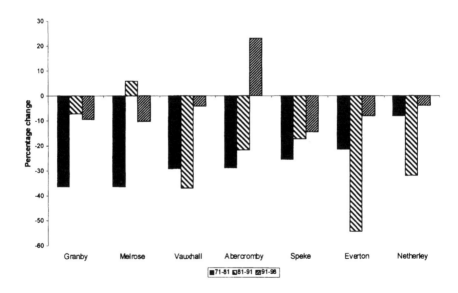

Figure 9.3   Population change in selected wards

Source: Liverpool City Council, 1993; Office for National Statistics,
                 2003

Overall the extent of regeneration is mixed. The Government's Index of Deprivation still records four wards in Liverpool as being within the top ten most deprived wards in the country (out of a total of over 8400 wards).

These are Speke (2nd), Everton (4th), Vauxhall (6th) and Granby (10th) (ODPM, 2000). This, in itself is powerful evidence of the failure to make significant progress against social and economic measures of regeneration. Nevertheless, there has been progress, particularly in Abercromby and parts of Granby where there has been strong evidence of private housing investment and some gentrification. As mentioned above these areas have received investment through a number of high-profile regeneration programmes. Much of the derelict former south docks that lay within Abercromby have been brought back into beneficial use through the activities of the MDC. City Challenge was successful in bringing substantial property investment to the area. The Ropewalks partnership appears to have been effective in kick-starting improvements to the public realm, bringing in new facilities, redeveloping derelict properties, refurbishing and upgrading others. Over many years Granby benefited from housing improvements through HAA and GIA investment and more recently from Estate Action schemes and the Granby Triangle Renewal Area. Thus in at least some parts of these wards there seems to be built up a momentum of investment that will hopefully continue into the future.

Vauxhall has also benefited from housing investment, although of a very different kind: the housing co-operative initiatives, local authority housing investment, the North Liverpool Partnership and LHAT. The MDC invested heavily in the infrastructure of 'Atlantic Avenue' and adjoining industrial areas. The north docks and their hinterland now form part of an important economic regeneration area.

Other areas dominated by social housing estates, both in the inner city and in peripheral areas, have remained deeply problematic despite the efforts of successive government initiatives. Thus, for example, although the Speke/Garston Partnership, Speke/Garston Development Company and SLH have shown success in improving the physical environment and attracting economic development, the scale of the social problems in these areas remain substantial. Other inner area wards such as Smithdown and Kensington have, until recently, seen less attention from regeneration policies and typify of a number of inner wards where problems of low housing demand and out migration were beginning to emerge as a serious problem in the late nineteen nineties. It is perhaps in this type of 'twilight area' where the greatest urban regeneration challenges of the coming decade will be found. The success of the New Deal for Communities regeneration programme in Kensington will be keenly watched.

Sustainable Development

Although debate about the environmental agenda had been developing through the previous two decades, a major breakthrough came in 1987 with the publication of the Bruntland Report, *Our Common Future* (World Commission on Environment and Development, 1987) with its definition of sustainable development 'meeting the needs of the present generation without compromising the chances of future generations to meet their needs'. The report was welcomed by the British government and sustainability became the guiding principle in a major rewriting of government policy towards the environment. With the publication of *This Common Inheritance*, launched by Chris Patten (Secretary of State for the Environment et al, 1990), environmental policy moved to centre stage in government thinking – at least for a while (Francis A, 2001).

The Department of the Environment produced guidelines for the environmental appraisal of development plans (Department of the Environment, 1993) and in 1994 a second document (*Sustainable Development: the UK Strategy*) recognised the importance of the planning system as a key instrument in delivering a sustainable pattern of urban land and the importance of urban regeneration in re-using previously urbanised areas in the most efficient way by making them more attractive places in which to live and work (DOE, 1994, p221). Given the Conservative government's previous anti-planning stance this was a remarkable turnaround. The pressure for sustainable development continued with policy being continually updated and improved. By 1998 the Labour government had issued a policy document: *Sustainable Regeneration: Good Practice Guide* (DTLR, 1998) in which it was stated that:

> The achievement of sustainable development is one of the most widely expressed aspirations of current public policy. Sustainability lies at the heart of a range of international and domestic programmes, and the goals of attaining sustainable development in economic and environmental fields attracts non-partisan support (DTLR, 1998, p1).

But the question for us here is the extent to which urban planning and regeneration in Liverpool embraced this modern environmental agenda. To what extent have plans and policies pursued nothing more than economic growth or have they sought to modify market behaviour for the sake of the environment? Figure 9.4 gives an indication of how environmental issues have been dealt with in plans for Liverpool and its regeneration over the past four decades. The criteria cover three areas of concern: *Global Sustainability* (atmospheric and climatic stability and the conservation of biodiversity); *Natural Resources* (husbanding, use and protection of air, water, land and mineral resources); and *Local Environmental Quality,* (conservation of local environments and systems ranging from open land to cultural heritage).

	Liverpool Interim Planning Policy Statement 1965 and associated policies	Merseyside Structure Plan 1979/ Current Planning Policies 1983	Liverpool Unitary Development 1996 and associated policies
Global sustainability			
Transport energy efficiency (reducing trips and proportion of motorised trips)	+	-	o
CO_2 emissions and absorption (e.g. increasing tree cover)	o	o	o
Natural resources			
Wildlife habitats (Safeguarding designated sites/ increasing wildlife potential)	o	+	+
Air and water quality (reducing pollutants)	o	+	+
Land and soil quality (reducing dereliction/contamination)	o	+	+
Local environmental quality			
Landscape and open land (enhancing landscape/ protecting open land)	o	o	+
Townscape (enhancing townscape)	+	o	o
Built and cultural heritage (safeguarding architectural and historic places and characteristics)	o	+	+
Maintenance of buildings and infrastructure (maintenance, repair and refurbishment)	+	+	o
Health, safety and amenity (improving personal health, safety / peace and quietude)	o	o	o

+ = positive effect   o = neutral or balanced effect   - = negative effect

Figure 9.4   Sustainable development in plans for Liverpool

Source: author's analysis

Although the plans and policies of the 1960s were published long before the emergence of the environmental agenda, planners had traditionally embraced a concern for 'amenity' and 'efficiency' in town design that naturally led them towards many of the goals of sustainable development.

Such traditional concerns included the control of urban sprawl, an efficient distribution of land uses, the appropriate provision of facilities and amenities, the avoidance of conflict between neighbouring land uses, and good urban design. Whilst the IPPS called for substantial road investment, the need to control the growth of traffic was recognised and public transport investments and co-ordination were being integrated into plans for the city's future development. Even though the importance of reducing CO^2 emissions was not understood, there was a strong desire to maintain and even increase the proportion of trips make by public transport in order to maximise the efficiency of the transport system.

This plan had little to say about the protection of natural resources within the city (other than open space for amenity and recreational purposes) and the reclamation of derelict and contaminated land had not yet become an issue. There was a strong element of physical determinism and urban design in plans from this period. The City Centre Plan in particular had a clear vision of the form and visual qualities of the future city centre. However, within this vision the protection of the existing built heritage had not developed beyond the preservation of individual buildings or groups of buildings of architectural or historic interest. There was little understanding of the idea of area conservation or the re-use of buildings because of the embedded previous investment they represented or as features of social or psychological importance to local people. On the other hand, since the publication of the Buchanan Report, the notion of 'environmental areas' protected from traffic intrusion and noise, had entered the theory of planning and was being implemented within the city.

The Inner Areas Plan Review gave some indication of the rapid evolution of planning theory in this period and how far the city planners had developed their thinking over the previous decade. Unemployment had become a serious issue and the City Council argued that environmental quality would be a factor in attracting inward investment.

> The Corporation is very conscious of the desirability of improving the physical environment...besides making the Inner Areas more attractive to industrialists who are considering expansion, has also provided work for more than 300 men engaged by the Corporation (Liverpool City Council, 1974, p12).

The review also included a number of other policies for environmental enhancements and took a hard and restrictive line against emerging proposals for out-of-centre retail developments, taking the view that hypermarkets and retail warehouses would be poorly served by public transport, generate congestion and erode the viability of existing centre (Liverpool City Council, 1974, p26).

In many ways this was a sophisticated plan that contained many of the environmentally positive policies that would be welcomed today: local environmental improvements; greening the city; conservation of built and natural heritage; concentration of major traffic generating activities at the most accessible points; investment and planning in support of public transport. But at the same time it is possible to perceive the beginnings of the emerging economic agenda with the unemployed and the encouragement of investment being given as reasons for the support of environmental improvements.

Three months after this plan was published the City Council as a county borough ceased to exist. Although it was not apparent at the time, looking back this was perhaps almost as significant for the future failures of urban policy as the impending economic crisis. Responsibilities for urban planning and regeneration were split between the new Merseyside County Council and the metropolitan district of Liverpool. Waste collection became a district function, waste disposal a county function. Public transport went to an agency of the county council and a number of environmental powers went with water supply and sewage disposal to a regional water authority. The opportunity for corporate planning and policy making, that had evidently begun to emerge in the city, was seriously eroded. The age of inter-corporate policy making had arrived.

The Merseyside Structure Plan strategy for urban regeneration would encourage investment in the inner urban areas, enhancing the environment and encouraging housing and economic expansion on derelict land whilst restricting peripheral development to a minimum. At the strategic level such an approach seemed to accord well with the principles of environmental sustainability: the city was to be kept as compact as possible, urban land was to be re-used and the urbanisation of open land would be kept to a minimum.

There were policies for environmental enhancement including reclaiming derelict land, landscaping of vacant land and planting trees in areas of need, safeguarding the external environment of residential and industrial areas, controlling industrial pollution, improving waste collection and disposal methods, refurbishing and re-using commercial and residential buildings and the conservation of areas of architectural, historic or archaeological interest. There was also a substantial array of environmental policies aimed at protecting natural resources and open land, many considered to be innovatory at the time.

However, in transportation planning some tension between environmental and economic aims could be observed. Policies included the provision of bus and rail services to give *reasonable* access to jobs, shops and other facilities at a *reasonable* fare, whilst improving the strategic highway network, to *attract* longer distance and heavy goods traffic (author's italics). This was less ambitious than the earlier proposals from the former City Council and less orientated towards environmental objectives. On the one hand public transport investment proposals had been reduced in response to cut backs in public expenditure yet selected proposals to build new strategic urban highways within the inner urban areas were retained and justified on the grounds of supporting economic growth.

Not all urban regeneration policies were contained within the Merseyside Structure Plan. Much spending was to be channelled through the Liverpool Inner City Partnership, with its aim of achieving the economic, social and environmental regeneration of inner Liverpool. However, with regard to the environment, the Partnership's approach was very clearly driven by economic considerations:

> There is an overwhelming need, at the present time, to upgrade substantially the physical environment to improve the image of the City and restore confidence in its future. It is now widely recognised that the existing poor environment has a fundamental rather than marginal effect on both residents and employers (Liverpool Inner Areas Partnership Committee, 1978, p5).

The role of public transport was clearly seen as marginal and there was no strategy to shift modal split in its favour. In contrast the inner ring road was to be given the highest priority in terms of improving the strategic highway network. Furthermore, it was accepted that there was:

> a need to keep the level of revenue support (for public transport) to a minimum (ibid, p74).

Between 1981 and 1998 the Merseyside Development Corporation pursued its own regeneration agenda independent of local planning control. After 1993 most urban regeneration activity was supported through the Single Regeneration Budget or English Partnerships (later the North West Development Agency). In appraising projects for financial assistance consideration was usually given to the strategic planning context, scale and practicality of the proposals as well as the degree of risk and most importantly the question of value for money. Whilst these criteria did not exclude sustainability it was certainly not a priority. The Single Regeneration Budget Challenge Fund was intended to promote regeneration programmes responsive to local needs and priorities. Although most SRB programmes within Liverpool contributed to overall improvements in the quality of urban life and therefore worked towards a more compact and sustainable city, in detail few of these programmes gave any significant attention to the new environmental agenda (Booth M, 1996, p86).

It was many years after the publication of the Merseyside Structure Plan before the city was to see the production of another statutory development plan for the city. During the eighties environmental planning was being undermined by central government through a series of high profile reversals of planning decisions on appeal, attempts to streamline the development plan system, and the introduction of weaker planning controls through the designation of Enterprise Zones and Simplified Planning Zones. By 1986 Merseyside County Council had been abolished and its powers divided between the metropolitan boroughs, central government agencies and joint boards.

Liverpool City Council became a unitary planning authority and was obliged to prepare a Unitary Development Plan (UDP). The overriding aim of the Liverpool UDP was urban regeneration. The main thrust of the plan was to encourage investment, particularly within the inner city and central area whilst restricting peripheral growth and protecting the natural and built heritage of the city. Although environmental considerations figured strongly in the plan, they were not its driving force. Sustainable development was interpreted as:

> sustainable development does not mean having less economic development, what it requires is that decisions throughout society are taken with proper regard to their environmental impact (Liverpool City Council, 1996, p35).

This was weaker that the Bruntland definition of sustainable development and seemed to indicate that economic development was at least as important to the City Council as the long term protection of the environment. Secondly, a primary objective of the UDP was to stem the decline in economic activity within the city, whereas the plan would merely *attempt to reconcile conflicts* between facilitating beneficial development and protecting and improving the local and wider environment (Liverpool City Council, 1996, p31). Furthermore, some key aspects of sustainable development, such as reductions in energy use and energy conservation were neither stated as strategic objectives nor given any particular priority.

On the positive side, the plan was one of the first UDP in the country to include an environmental appraisal of its policies. The plan went at least some way towards fulfilling most of the environmental criteria that might reasonably be applied to the planning of urban regeneration. The protection of natural resources within the locality, and the improvement of local environmental conditions were well addressed in the plan. There were strong policies for the protection of wildlife habitats and increasing tree cover through the Mersey Forest. Air, water, land and soil quality were all to be improved. Conservation of the city's built and cultural heritage had become a major priority. However, the plan remained weak in relation to transport policy and showed little inclination to seriously respond to issues affecting global sustainability.

Overall, a strong traditional concern for amenity and efficiency has made much planning theory and practice compatible with notions of sustainable development. However, although some progress has been made in recent years, especially in the protection of existing built and natural environments, planning and regeneration policies in Liverpool lag some way behind the best European practice terms of their contribution towards the modern environmental agenda. This is especially true with regard to the whole field of transportation planning.

Challenges to Urban Planning

Planning theory is concerned with a number of questions, firstly with the question of the justification for planning or intervention in the operation of free markets – why plan? Secondly, there is the question of how to plan? What are the mechanisms and tools of planning? How can the planning system be made more effective? Thirdly, there is a body of theory that is concerned with questions about what it is planning is trying to achieve? These theories in planning are concerned with determining legitimate aims and objectives for plans. One of the conclusions from this study of urban planning and regeneration in Liverpool over the past decades has to be that the idea of planning has changed and in some senses has become increasingly marginalized as society has turned towards the market for solutions to urban problems. As the planning process has become marginalized so it has become more important to debate these questions about the reasons, purposes and justifications for planning: theories 'for' planning. It could be suggested that the idea of 'planning' in Liverpool and in many other British cities is in crisis. Notions of competition, the survival of the fittest and rewards for the most successful can be found throughout much of urban policy. So the question arises: does this matter? Do these market-based approaches produce solutions that are better than those that could be produced through a planned approach? Or, do these solutions lead to social and environmental costs that could be avoided through a stronger planning process? Is it possible to justify a return to a more comprehensive planning system within Liverpool or are better results being produced through the market?

There have been many writers who have sought to justify or question planned interventions in the workings of the free-market. Pickvance (1977) argued that the determining factor in urban development was the operation of market forces and that these forces were, in reality, subject to very little constraint. He suggested that the powers of planners were essentially negative: the power to refuse planning applications. His point was that the development initiative has to come from developers and that planners merely modify the nature and location of that development. This is a powerful argument today but was not always so. Back in the early post-war years much, indeed most, development was initiated by the public sector. From the end of the war until the early 1950s most housebuilding was undertaken on behalf of councils or new town development corporations. The public sector rebuilt town centres and created new industrial estates, schools, hospitals and other community facilities. When the Town and Country Planning Act was passed in 1947 it was anticipated that the (positive) planning of public investment would be as important as the (negative) control of private development. In 1966 Liverpool City Council published a book: 'Liverpool Builds' in which they proudly recorded and displayed their building achievements in the post-war period (Bradbury, 1966). It is a substantial book containing a record of the many houses and flats, schools, clinics, libraries and other buildings that the council had provided for the city in accordance with its own plans and reflects a real sense of pride in civic achievement. The building projects described in the book illustrate the council's attempt to ensure a rational and efficient use of resources to achieve its aims. Aims that were, by and large, based upon the implementation of the principles of the welfare state.

Similarly, it is possible to look at the city council's plans of the 1960s and see massive intervention in the market. High value land near the central area was removed from the market and allocated for social housing. Slum clearance had a major effect in diminishing the size of the private rented housing sector and replacing it with social housing allocated outside the market. Valuable urban land was given over to school playing fields and public open space.

However, over the decades, less and less building in Liverpool has been initiated by the public sector. There is today very little public expenditure on the infrastructure of roads and parks and schools. The great majority of residential development and virtually all commercial and industrial investment is either privately funded, or undertaken by quasi-governmental agencies beyond local democratic control. Furthermore much of the land and property that was removed from the market in the post-war period had, by the end of the century, been returned to the private sector in response to the pressures from the property industry in its search for ever more sources of investment profit. Since the early eighties Liverpool has seen the sale of many council houses. Local authority housebuilding has virtually ceased as growing reliance is placed upon private housing development to maintain supply. Some playing fields have been sold off for more profitable uses. Many public services have been privatised and within this process their assets (the land and property owned by the former nationalised industries and public utilities) have passed back into private ownership where future uses will be determined by market forces rather than public policy.

In these circumstances the city council may produce development plans: not to guide its own investment but to control investment by others. Today it is private investors and quasi-governmental agencies that initiate and promote the location and scale of development. Anticipating the location, nature and timing of private investment is extremely difficult. On many occasions in the past, planning forecasts have been wrong and frequently plans have had to be modified to reflect market changes. It can be argued that the effectiveness of development plans have been undermined because of their failure to accurately anticipate strategic economic and investment trends. Planners have been placed in a situation of reacting to rather than dictating the nature and location of investment.

Some examples illustrate the point. The IPPS of 1965 and the MALTS were predicated upon the premiss of accommodating and planning for continued economic growth within the region. Within only a few years of the publication of these plans their forecasts for local economic and population growth, which underpinned the proposed highway, public transport, housing, industrial and commercial investment were shown to be not only wrong in scale but in direction as well. Some years later, there was nothing in existing development plans to suggest that the South Docks would become vacant and should be redeveloped in a particular way. Indeed, even the proposals contained within the MDC's own 'master plan', based upon the simple notion of achieving a beneficial use on these sites, underwent substantial modification and adjustment to meet the needs of the property market. Thirdly, there is the recent phenomenon of new housing investment in Liverpool city centre. Housing investment in the centre grew from virtually nothing in the 1995 into a significant and vibrant housing sub-market by the year 2000. This shift in the location of housing investment was not anticipated in any development plan. This surge is not the consequence of positive planning decisions taken by the city council but of market forces that have fortuitously coincided with the aims of the city council and have therefore been encouraged and facilitated. Finally the recent growth of Liverpool Airport was not planned. Its growth was not directed by any strategy but by a growth in demand for air travel, changes in the structure of the air transport industry and the investment decisions of individual airlines. These changes stimulated investment in the airport and fortuitously, other local economic benefits.

There is little in these arguments to suggest that the declining importance of plan making and development plans is of much consequence to the future of Liverpool. On the other hand, the strength of economic demand may be an important factor in determining the ability of the planning system to influence and shape development. It has been argued that in circumstances of weak economic demand local governments are grateful for any investment and feel constrained not to deter development with heavy regulations or tax impediments for fear it will transfer elsewhere. Conversely, when demand for property development is buoyant, local governments feel more able to extract some of the surplus profit from development in the form of taxation, improvements in design quality or the provision of additional facilities. According to Rydin:

> Less pressure on local authorities for planning permission for development schemes (leads to) less possibility of regulating that development pressure to meet regeneration objectives and fewer possibilities for planning gain (Rydin, 1998, p195).

It is clear that the demand for property development in Liverpool has varied over time broadly in line with national investment cycles. However, demand in Liverpool, especially for commercial and retail development, has been consistently weaker than the national average throughout most of the period under consideration. It is very difficult to test the proposition that differences in economic circumstances over time have affected planning and urban policy decisions in Liverpool. However, it is possible to argue that the relative weakness of the local economy compared with other UK cities has weakened the regulatory and negotiating power of the planning system in Liverpool compared with those other cities. This can be demonstrated by comparing the limited amount of 'planning gain' achieved in Liverpool with the almost commonplace expectation of planning gain in the South-East; or the quality of the public realm in prosperous cities centres such as Bristol, Bath, Chester and even Manchester with the relative inadequacy of such spaces in Liverpool.

Others would argue that planning has become little more than a tool of the capitalist system: that the purpose of planning is to create and maintain the best conditions for private property investment. Many writers, particularly in the seventies, viewed town planning in this light (see for example, Harvey, 1973). It is easy to see how such a view could gain support. In the post war boom years, when even Liverpool experienced a wave of relative prosperity, the aims of planning seemed to be somewhat detached or separate from the needs of the economic system. In the IPPS, Alderman Sefton wrote about 'modernising the city'. Admittedly he was in part seeking to provide the infrastructure for a modern economy but there also appeared to be a more altruistic motive concerned with the welfare of the citizens and the quality of the environment, not for economic ends but for their own sakes. However after the 1973 recession the tone changed. The economy was in decline and regeneration had become the principle aim of urban policy. In these circumstances the apparatus of planning was swung behind the need to create 'much needed jobs'. Merseyside Council Council's Stage One Report stated that:

> the basic planning powers of development control and land allocation, designed to control new development where the market pressure was assured, *are less important than* policies which initiate redevelopment and regeneration (Merseyside County Council, 1975, p1) (author's emphasis).

With the arrival of Thatcherism the message became even stronger. The purpose of the Merseyside Development Corporation was to promote regeneration through the reclamation of derelict land and buildings. To facilitate this process the MDC was given substantial powers of development control that effectively made it the planning authority for its area. In other words within its designated area planning powers were exercised by an agency whose principle aim was to stimulate property investment.

There are other examples. The planning system across the country had successfully resisted major out of town retail developments until the relaxation of planning powers within EZ allowed a number of major regional shopping centres and retail parks, including an example of the latter within the Speke Enterprise Zone. Through the latter years of the century it became the planners' task to provide the infrastructure to underpin the latest property investment whim or fad. Not in Liverpool fortunately but elsewhere, such as in Chester, agricultural land was given over to a business park on the basis that it created employment with little regard to the consequences for urban sprawl or traffic generation. Investment in major infrastructure projects such as airport developments, inter-urban highways and power stations almost seemed to go ahead regardless of the planning system.

So there are strong reasons for arguing that the planning system has done little more than support economic development and in particular, facilitate property investment. And yet this is not actually the case. In fact there is a real and constant battle between the desire of landowners and developers to seek the maximum profit from land regardless of the social costs of development, and the planning system that seeks to minimise social costs and maximise social benefit. One of the major triumphs of planning is that it has and continues to stand in opposition to unfettered property investment. This can be seen in Liverpool in slum clearance, transport policy, the green belt, urban regeneration and more recently in the impact of the environmental movement.

The Importance of Urban Planning

In 1961 nearly 40 per cent of the city's housing stock was without exclusive use of all amenities: inside toilet, hot and cold running water, etc. These dwellings were unfit for human habitation. In time the market would have dealt with the problem: if the value of these dwellings fell below the value of the land upon which they were built, they would be cleared and replaced by new developments. In a city like Liverpool, with a sluggish local economy, this redevelopment process could take many years, leaving inhabitants living in slums housing for long periods of time. It would have been cheaper for the state not to intervene but to let this process take its course. However, the political pressures from the population were such that the state did take action to clear unfit dwellings and replace them with new housing faster than the market would have done if left to its own devices. This intervention had a number of benefits: it had social benefits in meeting housing need and mollifying a potentially dissatisfied electorate; it had physical benefits that led to improvements in public (as well as individual) health; and it had economic benefits in stimulating demand for house-building, which was welcomed by the construction industry. In this win-win situation planned intervention by the state, through the city council, was on a massive scale. Between 1961 and 1966 an average of over 1,250 dwellings were cleared and replaced each year. Between 1967 and 1974 this rose to over 2800 dwellings a year, peaking at 4,239 clearances in 1973. By this time the worst slums had been tackled and a crude housing surplus achieved. Public support for clearance declined amidst a growing body of adverse experiences of the re-housing process and the poor quality of some replacement housing. Political support ebbed away sharply and by the late seventies the average annual rate of clearance and rebuilding had fallen back below 1,000. By the 1980s housing clearance had ceased to be a significant element in urban policy.

The problems of urban transport had in a similar fashion come to command substantial political consideration by the middle of the 1960s. With rising car ownership, congestion and road accidents, greater state intervention in the regulation and planning of the transport system was becoming essential. Through MALTS and the IPPS major highway and public transport investments were planned for the city. As with housing renewal these transport proposals offered a major physical restructuring of the city with the benefits of concentrating traffic along defined corridors, creating environmental areas elsewhere, improving accessibility, reducing travel times, and giving huge economic benefits to the construction and civil engineering industries. There were many reasons for the private sector to support these changes. However, a number of factors mitigated against the implementation of these bold policies. Firstly, the socio-political benefits were limited. There was no great social movement for better transport as there had been for housing and it was difficult politically to justify increased state expenditure on public transport at a time of falling bus and rail patronage. Secondly, whilst the economic benefits to the construction sector of road and railway building were substantial, those to other economic sectors were more difficult to define. The benefits to industry in terms of travel time and cost savings from state investment in inter-regional motorways were many times greater than those that could be achieved from urban motorway building. Futhermore, the actual financial cost of urban motorway schemes was proving to be far higher than anticipated and impossible to support. By the 1970s, pressures to reduce public spending, combined with a new form of pressure from the emerging environmental lobby, pushed both central and local government into abandoning many of these planned transport improvements. In these circumstances expensive urban road proposals proved no more durable than public transport schemes. By 1975 the grandiose proposals for the 'Middle North-South Primary' and the 'Third River Crossing' had been abandoned, as had the 'Outer Rail Loop' (now a recreational footpath) and a number of other more prosaic proposals for local rail passenger improvements. It has only been in the last decade that a coincidence of interests between the economic development and environmental lobbies has led to some modest investments in new stations (for example, Brunswick and Wavertree Technology Park), bus priority lanes and traffic management schemes. Political support for major urban road building has evaporated completely.

Under market conditions urban areas have tended to expand outwards. Firms have tended to suburbanise as they seek larger sites on cheaper land to support changing production methods and better access to the inter-regional motorway network. Households have tended to suburbanise as increased real incomes have allowed them to purchase greater quantities of housing space (usually on larger plots) and to support greater personal transport costs. Historically much of the slack within inner urban areas was taken up by the expansion of central business districts into the neighbouring 'zone of transition', and by lower income households filtering up through the housing system. However, by the early 1970s in Liverpool and other large cities in the UK, the rate of suburbanisation was outstripping the rate at which replacement land uses could occupy the vacated premises. Additional factors such as international competition and technological change were also causing the closure of many other firms within inner urban areas. Social tensions, such as those between indigenous and immigrant communities, unemployment, rising crime rates, and deteriorating urban environments were further encouraging a 'flight to the suburbs' by those households who could afford the move.

In these circumstances in an unfettered urban economy, the tendency would be for inner city industrial and residential areas to become abandoned and derelict. Only when land values reached rock-bottom would 'pioneer' developers gradually creep back in to re-colonise abandoned areas of the city. Examples of this process can be found in many cities of the United States, especially in the north-eastern 'rust-belt' cities such as Baltimore or Detroit. But in the UK this market process was never allowed to progress unfettered, and Liverpool was very much in the vanguard of developing policy responses to control such trends: urban regeneration coupled with restrictions on peripheral growth.

The idea of 'green belts' around cities dates from the inter-war era when suburbanisation and urban sprawl were at their most rapid. (The built up area of Liverpool doubled in the twenty years between 1919 and 1939). But despite the concern it was to be many years before an effective 'green belt' was put in place. The emergence of the inner city problem added a second reason to restrict peripheral growth and the Merseyside Green Belt Local Plan was finally approved in 1983. Since that time the 'green belt' (most of which is not in Liverpool but in adjoining authorities) has held up remarkably well with few incursions by urban development.   Part of the explanation lies within the economics of the housing market. Most housing in the UK is nowadays developed by private house-builders. Most of this output is produced by a small number of larger firms operating nationally and internationally. Relative to some other regions, notably the South-East, profit margins are somewhat lower in Merseyside, firms are less keen to develop here and consequently put less pressure on the 'green belt'. Nevertheless, it is still fair to say that the local planning system has been quite strong in defending the periphery from development. Between 1985 and 1997 a total of 3,690 hectares of land in Merseyside were developed for urban uses and of these only 900 hectares (24 per cent) were formerly in rural uses. That is to say, more than three-quarters of all development in Merseyside over the period was on existing urban land. This picture is reinforced by DTLR Land Use Change Statistics that show that on average over the same period nearly 70 per cent of all new housing in Merseyside was built on previously developed land, i.e. contributed to urban regeneration.

There are a number of reasons why policies for urban regeneration emerged. These can be divided into social and economic factors. On the social side one reason was simply the continuation of a tradition of intervention in slum housing (the sanitary movement) dating back to the 19th century. Although slum clearance was being wound down by the 1970s, the alternative policy, area improvement, required city council officers to engage in a process of community participation and to respond to their needs by providing social amenities and environmental improvements as well as the renovation of the dwelling stock. Liverpool was amongst the first cities to declare a General Improvement Area and was to become enthusiastic in its embrace of the new approach. A second reason was that the 1968 Town and Country Planning Act required local planning authorities to survey *social and economic conditions* in preparing their new structure plans. This was the first time that legislation had placed any such burden upon local councils. The effect was to make planners more aware than they ever had been before of differences in living conditions between different social groups and different spatial areas.

Again, in the field of social area analysis, Liverpool City Council was at the forefront of policy innovation, even if the impact of local government reorganisation delayed the preparation of the structure plan itself. A third reason was the real or perceived threat of social unrest. It was this fear that had led the Labour government in 1968 to introduce the Urban Programme. From this programme came the Community Development Projects and the

Inner Areas Studies that helped explain to central and local governments the true nature and costs of urban deprivation as well as the form that might be taken by policy intervention. Liverpool was an early recipient of Urban Programme funds as well as hosting the Vauxhall CDP, the Shelter Neighbourhood Action Project and one of the three Inner Areas Studies. Thus the city, its people and politicians made a major contribution to the development of modern urban regeneration policy.

On the economic side the arguments for intervention in urban regeneration were heavily influenced by the property industry's concern with maintaining land prices and the 'waste' of valuable vacant and derelict urban land. Industrialists were also able to argue to government that, however much they would like to invest in the inner urban areas, they simply could not afford to do so unless the government provided subsidies. Furthermore, the residual populations of these areas would not be able to access the jobs that would be created as they lacked the necessary skills. If the government wanted to see these populations re-employed it would have to pay for their education and training. Whilst the Labour government recognised the strength of these arguments in its 1977 White Paper: Policy for the Inner Cities, it was left mainly to the incoming Conservative government to implement policy. Under the 1980 Local Government, Planning and Land Act, local authorities were required to establish registers of all surplus land owned by public bodies, with a view to bringing the availability of such land to the attention of potential developers. The Secretary of State could direct the disposal of sites that were deemed to be 'unreasonably' withheld by the public sector owner. More importantly the same Act established the Urban Development Corporations with their massive powers to bring the largest (and potentially very valuable) areas of urban dereliction back into beneficial economic use. Enterprise Zones were designed to encourage economic development in derelict and deprived areas through a combination of subsidies and a relaxation of planning controls. In terms of planning theory these are curious policies for they seek to further the aims of planning policy (to regenerate urban areas) through mechanisms that tended to weaken local planning controls and the powers of local government. Yet they are a form of intervention that seeks to modify market behaviour and reduce social costs. Once more Liverpool was at the forefront of policy innovation: this time as the (initially unwilling) recipient of the Merseyside Development Corporation and the Speke Enterprise Zone.

Also during the same decade a revamped Urban Programme provided funds to subsidise local environmental improvements, not least to improve the 'image' of areas as a stimulus to inward investment, and to provide skills training for local people. By 1982 gap funding had arrived in the form of Urban Development Grants (later City Grant). This very precise and effective policy provided a cash subsidy to specific developers to enable them to cover the gap between the high cost of development and the low value of the completed premises in deprived areas. Government departments developed an array of other subsidies and initiatives during the decade whilst at the same time taking an ever greater role in direct action themselves, rather than relying on the agency of local government to implement policy. This in itself is an indication of how seriously the government was taking the problem of urban social and economic breakdown and the need for urban regeneration.

By 1990 Michael Heseltine had decided that more responsibility for urban regeneration could be transferred back to the local authorities provided they involved local communities and (significantly) businesses, in regeneration 'partnerships' and that they competed for funding from central government. This allowed central government and to a lesser extent, local business, to retain control over urban regeneration policy, whilst transferring much of the responsibility for policy development and implementation back to the local authorities. The first policy to be developed under this new philosophy was City Challenge. Liverpool was amongst the first round 'winners' of City Challenge funding with its City Centre East proposals. By 1993 central government had streamlined its approach to urban regeneration but the ideas behind City Challenge were refined and continued as the Single Regeneration Budget Challenge Fund. This policy, with its strong emphasis on competition between areas for limited funds and partnership between local government, communities and the private sector, remained in place until the end of the decade.

It could be argued that urban regeneration has been the single most important urban planning policy of the last thirty years. It has been a very significant intervention into the working of urban economies: land markets; labour markets; housing markets; and even capital markets. And yet its very importance has led to central government taking control of the policy and usurping local democratic and planning control in the 'national' interest. At the same time the private sector has perceived the policy to be of such importance that they also needed to become part of the control mechanism and to influence policy through the guise of 'partnerships' with local communities.

The growth of the environmental movement can be traced back at least as far as the work of the 'Club of Rome' (Meadows, 1972) but it was really after the Bruntland Report in 1987 that governments began to treat seriously the issue of sustainable development. Over the past decade since the heady days of the Rio Summit in 1992, interpretation of the concept has subtly evolved and changed. By 1994 the UK government was suggesting that sustainable development did not mean having less economic development but, on the contrary, a healthy economy would be better able to meet people's needs, and that new investment and environmental improvement could often go hand in hand (Department of the Environment, 1994, p7). This attempt at a political compromise between economic development and environmental sustainability was already at odds with earlier definitions of the concept. By the millennium the government had moved yet further with an ambiguous statement that 'sustainable development is about ensuring a better quality of life for everyone, now and for generations to come' (Department of the Environment, Transport and the Regions, 2000). There remained a similar ambiguity at the local level, for whilst Liverpool City Council signed up to Local Agenda 21, the primary objective of the UDP was to stimulate economic development.

The effect of this ambivalence towards sustainable development at both national and local level can be seen in the outcomes of policy. As has been discussed above, the city has been quite successful in regenerating the urban area in the sense encouraging property investment within the city and maintaining property values. This has been assisted by the implementation of strong 'green belt' policies. The outcomes of both of these polices, the retention of population and economic activity within the existing urban area and reduction in urban sprawl are, by coincidence, compatible with sustainable development, even if this was not their primary motivation. British planning, and Liverpool planning in particular, has historically been closely wedded to the notion of conserving the amenity, character and heritage to be found within the existing built environment. Again this is a field in which the city can claim notably achievements that are compatible with sustainable development. However, when it comes to the consideration of ecological issues and the conservation of natural resources the achievements of the city council, both as planning authority and as a corporate body, are much more modest. But perhaps the biggest failure so far has been in the field of energy conservation, particularly the failure to reduce the number and length of transport trips and the failure to shift modal split away from the motor vehicle. Both at national and local level transport planning compares poorly against that of almost any other major European country or continental city. It is in the field of transport planning more than anywhere else that policy must be improved, regulations tightened and investment increased if the city is to provide the attractive, efficient and sustainable environment, necessary for the 21st century.

Within Liverpool urban planning and regeneration, which was once almost uniquely the responsibility of the city council, has today been divided between a multitude of agencies ranging from private developers and privatised utilities companies, rail and bus operators, through government agencies and registered social landlords to short-life quasi-public regeneration partnerships. Each agency has its own agenda, its own management system, its own legal framework and its own priorities. The key to implementing planning policy in this situation is the ability to negotiate, to search for synergies and to build coalitions and partnerships for action. Unfortunately, just as partners in coalition governments have to forfeit some of their political agenda, so in urban planning, agencies have little choice but to make compromises that weaken their ability to achieve their own objectives (Couch and Dennemann, 2000). Furthermore, with the city council itself torn between economic and environmental objectives but more concerned with the former it seems unlikely that future development will be in any sense 'sustainable' except on those few occasions when there might be synergy between short term economic gain and long term environmental benefit.

So, over the last thirty years in Liverpool there has been a movement away from comprehensive planning towards sectoral policy making and ad hoc interventions. There has been a fragmentation also in the delivery of services, particularly in housing, health, education and transport. There has been a continuing process of privatisation and commodification of services. Nationally, economic activity continues to concentrate in the south. In the North West, Manchester has emerged as the dominant city and the Liverpool economy continues to decline relative to the region as a whole.

At the regional and local levels considerations of economic development appear to dominate decision-making and there are unresolved debates about the nature of governance and planning, especially at the regional level. Central government appears to be calling for stronger and more innovative urban regeneration policies yet at the same time it is critical of the traditional role of the planning process. It is difficult for planners to know how to react to these signals. The statutory planning process is stigmatised and marginalized, yet the work that planners actually do is valued. In the field of urban regeneration and in the struggle for sustainable development planners are at the forefront: shaping infrastruture investment; facilitating and controlling property development; designing solutions to urban and environmental problems; building partnerships for action; assembling funding packages; negotiating implementation mechanisms and monitoring progress. But too much of this activity takes place in an ad-hoc manner outside of any strategic framework. No successful major company makes decisions in such a way, with so little regard to consequences and externalities, so why should our towns and cities be subjected to such inadequacies. This is not the place to define a new planning agenda. But there is an undoubted need for a more critical appraisal of planning practice, which might start with a look back to the beginnings of comprehensive planning in the city, when it was linked to objectives of environmentalism, equity and social justice. To reaffirm the importance of a comprehensive urban planning system and to resolve the tensions between economic development and sustainable development are among the key challenges facing the city at the dawn of the 21st century.

Bibliography

Abel-Smith B and Townsend P, (1965), *The Poor and the Poorest,* Bell, London.

Abercrombie P, (1945), *Greater London Plan 1944*, HMSO, London.

Anderson B L and Stoney P J M, (eds), (1983), *Commerce, Industry and Transport: Studies in economic change on Merseyside,* Liverpool University Press, Liverpool.

Arnstein S, (1969), A ladder of citizen participation, *Journal of the American Institute of Planners*, Vol. 35, No. 4 pp. 216-224.

Atkinson R and Moon G, (1994), *Urban Policy in Britain*, Macmillan, London.

Audit Commission, (1989), *Urban Regeneration and Economic Development*, HMSO, London.

Ben-Tovim G, (1988), Race, politics and urban regeneration, in *Regenerating the Cities: the UK crisis and the US experience*, Parkinson M, Foley B and Judd D (Eds), Manchester University Press, Manchester.

Blowers A, (Ed), (1997), *Planning for a sustainable environment: a report by the Town and Country Planning Association*, Earthscan, London.

Bradbury R, (1966), *Liverpool Builds: 1945-1965*, Liverpool City Council.

Brauner S, (2000), Regeneration without confrontation, *Regeneration and Renewal*, Haymarket Business Publications, 20th October 2000.

Cabinet Office, (1988), *Action for Cities*, HMSO, London.

Central Advisory Council for Education, (1967), *Children and their Primary Schools* (The Plowden Report), HMSO, London.

Central Office of Information, (1981), *Merseyside*, Press Release, London.

Centre for Local Economic Strategies, (1990), *First Year Report of the CLES Monitoring Project on UDCs,* CLES, Manchester.

Chadwick G, (1971), *A Systems View of Planning*, Pergamon, Oxford.

City Centre Planning Group, (1965), *Liverpool City Centre Plan,* Liverpool City Council.

Commission for the European Communities, (1994), *Single Programming Document for Merseyside, 1994-1999*, CEC, Brussels.

Commission for the European Communities, (2000), *Single Programming Document for Merseyside, 2000-2006*, CEC, Brussels.

Cornfoot T, (1982), The Economy of Merseyside, in Gould W T S and Hodgkiss A G (Eds), *The Resources of Merseyside* Liverpool University Press, Liverpool.

Couch C, (1999), Housing Development in the City Centre, in *Planning Practice and Research,* Vol.14, No.1, pp. 69-86.

Couch C, (2000), Urban Renewal and Grants, in Allmendinger P, Prior A and Raemaekers J (Eds), *Introduction to Planning Practice*, Wiley, Chichester.

Couch C, (2000a), Community Based Housing Development: the example of the Liverpool Housing Action Trust, in *Environments by Design*, Vol.3, No.2, pp. 59-76.

Couch C, (2001), Urban Regeneration and the Environment in Liverpool: from the 1970s to the 1990s, in Couch C and Boemer H (Eds), *Economic Restructuring, Urban Change and Policy in the Ruhr and Merseyside, 1978-1998,* Anglo-German Foundation, London.

Couch C, Eva D and Lipscombe A, (2000), Renewal Areas in North West England, in *Planning Practice and Research,* Vol. 15, No. 3, pp. 257-267.

Couch C and Wynne S, (1986), *Housing Trends in Liverpool,* Liverpool Council for Voluntary Service, Liverpool.

Cullen G, (1961), *Townscape,* The Architectural Press, London.

Daniels P W, (1983), Merseyside's service and office economy: The key to future stability?, in Anderson and Stoney, op cit.

Department of the Environment, (1977), *Change or Decay; Final Report of the Liverpool Inner Area Study,* HMSO, London.

Department of the Environment, (1988), Planning Policy Guidance Note 1 *'Strategic Planning Guidance for Merseyside',* HMSO, London.

Department of the Environment, (1988), *Urban Land Markets in the United Kingdom,* HMSO, London.

Department of the Environment, (1992), *Planning Policy Guidance Note 12: Development Plans and Regional Planning Guidance,* HMSO, London.

Department of the Environment, (1993), *Environmental Appraisal of Development Plans: A Good Practice Guide,* HMSO, London.

Department of the Environment, (1994), *Assessing the Impact of Urban Policy,* HMSO, London.

Department of the Environment, (1994), *Sustainable Development: the UK Strategy,* HMSO, London.

Department of the Environment, (1996), *City Challenge, Interim National Evaluation,* The Stationery Office, London.

Derbyshire A, (2001), *Impact of Urban Policy in the Regeneration of the Granby Triangle,* unpublished masters dissertation, Liverpool John Moores University.

Docklands Action Group, (1975), *Liverpool South Docks: The Future,* c/o Liverpool University Settlement, Liverpool.

DTZ Pieda, 2000, *Englands North West: a Strategy Towards 2020 and Regional Planning Guidance Review - Spatial Implications of the North West Regional Strategy,* DETR, London.

Francis A, (2001), Environmental Planning in Britain, 1978-1997: sustainable optimism? In Couch C and Boemer H, op cit.

Gibson M and Langstaff M, (1982), *An Introduction to Urban Renewal,* Hutchinson, London.

Gifford, (1989), *Loosen the Shackles,* The Gifford Inquiry, Karia Press, Liverpool.

Gillespie C, (1998), *Single Regeneration Budget (Challenge Fund) – Is it Targeted?* unpublished masters dissertation, Liverpool John Moores University.

Gilman S and Burn S, (1994), Dockland activities: technology and change, in Gould W T S and Hodgkiss A (Eds), *The Resources of Merseyside,* Liverpool University Press, Liverpool.

Harvey D, (1973), *Social Justice and the City,* Edward Arnold, London.

Hayes M, (1987), *Merseyside Development Corporation: The Liverpool Experience*, Liverpool City Council.

House of Commons Environment Committee, (1983), *The problems of Management of Urban Renewal (Appraisal of the Recent Initiatives in Merseyside)*, HMSO, London.

Hyde F, (1971), *Liverpool and the Mersey: the development of a port 1700 – 1970*, David and Charles, Newton Abbot.

Jacobs J, (1961), *The Death and Life of Great American Cities*, Random House, New York.

Labour Market Strategy Group, (1995), *Merseyside Economic Assessment*, Government Office for Merseyside, Liverpool.

Lawless P, (1988), *Britain's Inner Cities*, Harper and Row, London.

Liverpool City Challenge, (1991), *City Challenge: Liverpool City Centre East, Preliminary Submission*, Liverpool City Council.

Liverpool City Council, (1965), *Liverpool Interim Planning Policy Statement*, Liverpool.

Liverpool City Council and others, (1969), *Merseyside Area Land Use / Transportation Study*, Liverpool.

Liverpool City Council, (1974), *Liverpool Inner Areas Plan Review*, Liverpool.

Liverpool City Council, (1983), *Renewing Liverpool's Council Housing: An Overall Strategy for the 1980s*, Liverpool.

Liverpool City Council, (1987), *Past Trends, Future Prospects*, Liverpool.

Liverpool City Council, (1987), *City Centre Strategy Review*, Liverpool.

Liverpool City Council, (1993), *City Centre Plan*, Liverpool.

Liverpool City Council and NURAS, (1995), *Renewing Granby* (newsletter) 3rd edition, Liverpool.

Liverpool City Council, (1996), *Liverpool Unitary Development Plan*, (Deposit Draft), Liverpool.

Liverpool City Planning Department, (1966), *Social Survey: A Study of the Inner Areas of Liverpool*, Liverpool City Council, Liverpool.

Liverpool City Planning Department, (1970), *Inner Areas Plan*, Liverpool City Council, Liverpool.

Liverpool City Planning Department, (1970), *Social Malaise in Liverpool*, Liverpool City Council, Liverpool.

Liverpool City Planning Department, (1972), *Liverpool South Docks: Principles of Redevelopment*, Liverpool City Council, Liverpool.

Liverpool City Planning Department, (1983), *Current Planning Policies and Development Programmes*, Liverpool City Council, Liverpool.

Liverpool Inner City Partnership Committee, (1978), *Liverpool Inner City Partnership Programme 1979-82*, Liverpool.

Lloyd P E, (1970), The Impact of Development on Merseyside, in Lawton R and Cunningham C M (Eds), *Merseyside: Social and Economic Studies*, Longman, London.

Loney M, (1983), *Community Against Government: The British Community Development Project 1968-1978*, Heinemann, London.

Longstreth-Thompson P, (1945), *A Regional Plan for Merseyside*, HMSO, London.

McIntyre B, (1995), *The Effectiveness of Urban Policy, a Case Study: The Liverpool Inner City Ward of Vauxhall*, unpublished masters dissertation, Liverpool John Moores University.

McLoughlin J B, (1969), *Urban and Regional Planning: A Systems Approach*, Faber and Faber, London.

Marriner S, (1982), *The Economic and Social Development of Merseyside*, Croom Helm, London.

Meegan R, (1993), Urban Development Corporations, Urban Entrepreneurialism and Locality, in Imrie R and Thomas H (Eds), *British Urban Policy and the Urban Development Corporations*, Paul Chapman Publishing, London.

Merseyside County Council, (1974), *Consultation*, Liverpool.

Merseyside County Council, (1975), *Stage One Report*, Liverpool.

Merseyside County Council, (1979), *South Docks Prospectus*, Liverpool.

Merseyside County Council, (1979), *Merseyside Structure Plan*, Liverpool.

Merseyside County Council, (1983), *Merseyside Green Belt Local Plan*, Liverpool.

Merseyside County Council, (1985), *Agenda for Merseyside*, Liverpool.

Merseyside Development Corporation, (1981), *Initial Development Strategy*, Liverpool.

Merseyside Development Corporation, (1990a), *Development Strategy*, Liverpool.

Merseyside Development Corporation, (1990b), *Liverpool Waterfront Area Strategy*, Liverpool.

Merseyside Development Corporation, (1990c), *South Liverpool Area Strategy*, Liverpool.

Merseyside Development Corporation, (1997), *Annual Report 1996/97*, Liverpool.

Merseyside Planning Officers Group on Structure Planning, (1972), *Merseyside: A Review*, Liverpool.

Merseyside Structure Plan Team, (1973), *Merseyside, 1986: if present trends and policies continue*, Liverpool.

Merseyside Task Force, (1988), *Merseyside Integrated Development Operation 1988-92*, Liverpool.

Midwinter E, (1972), *Priority Education: An Account of the Liverpool Project*, Penguin, Harmondsworth.

Ministry of Housing and Local Government (MHLG), (1962), *Town Centres, Approach to Renewal*, HMSO, London.

Ministry of Housing and Local Government (MHLG), (1963), *Traffic in Towns* (The Buchanan Report), HMSO, London.

Ministry of Housing and Local Government (MHLG), (1965), *The Future of Development Plans*, HMSO, London.

Ministry of Housing and Local Government (MHLG), (1967), *Old Houses into New Homes* (Cmnd 3602) HMSO, London.

Mitchell and Rapkin, (1953), *Urban Traffic: a function of land use*, Columbia University Press, New York.

Muchnick D M, (1970), *Urban Renewal in Liverpool: A Study in the Politics of Redevelopment*, Occasional Papers on Social Administration 33, Bell and sons, London.

Nairn I, (1955), *Outrage*, The Architectural Press, London.

North West Development Agency, (2000), *A Strategy Towards 2020*, NWDA, Warrington.

North West Joint Planning Team, (1974), *Strategic Plan for the North West,* HMSO, London.

North West Regional Assembly, (2000), *People, Places and Prosperity: Draft Regional Planning Guidance for the North West*, NWRA, Wigan.

Office of the Deputy Prime Minister, (2000), *Indices of Deprivation 2000*, ODPM Housing Support Unit, London.

PA Cambridge Economic Consultants, (1987), *An Evaluation of the Enterprise Zone Experiment*, HMSO, London.

Parkinson M, (1985), *Liverpool on the Brink*, Policy Journals, Hermitage.

Parkinson M, (1988), Liverpool's fiscal crisis: an anatomy of failure, in *Regenerating the Cities: the UK crisis and the US experience*, Parkinson M, Foley B and Judd D (Eds), Manchester University Press, Manchester.

Parkinson M, (1990), Leadership and Regeneration in Liverpool: Confusion, Confrontation, or Coalition? in Judd D and Parkinson M (Eds), *Leadership and Urban Regeneration,* Sage Publications, London.

Pickvance C, (1977), Physical planning and market forces in urban development, *National Westminster Bank Quarterly Review* (Quoted in Taylor, op cit).

Planning Research Applications Group (PRAG), (1975), Liverpool Social Area Analysis (Interim Report), Centre for Environmental Studies, London.

Regional Planning Guidance Sub-Group, (1994), *Greener Growth,* North West Regional Association, Manchester.

Robson B, (1988), *Those Inner Cities*, Clarendon Press, Oxford.

Robson B and Parkinson M, (2000), *Urban Regeneration Companies: A Process Evaluation*, Department of the Environment, Transport and the Regions, London.

Rydin Y, (1998), *Urban and Environmental Planning in the UK,* Macmillan, London.

Shankland G, (1963), *Planning Consultants' Report No.8 Central Area Redevelopment: Two New Centres*, Liverpool City Council, Liverpool.

Shelter Neighbourhood Action Project, (1972), *Another Chance for Cities*, Shelter, London.

Shore P, (1976), *Inner Urban Policy – A Speech*, Central Office of Information, Press Notice 17th September, London.

Stoney P J M, (1983), The Port of Liverpool and the regional economy in the twentieth century, in Anderson and Stoney, op cit.

Taylor N, (1998), *Urban Planning Theory since 1945*, Sage Publications, London.

Topping P and Smith G, (1977), *Government Against Poverty? Liverpool Community Development Project 1970-75*, Social Evaluation Unit, University of Oxford.

Urban Design Quarterly, (1999), Case Study: Liverpool Vision, *Urban Design Quarterly*, No.76.

Urban Task Force (UTF), (1999), *Towards an Urban Renaissance*, E and F N Spon, London.

Young M and Willmott P, (1957), *Family and Kinship in East London,* Routledge and Kegan Paul, London.

Ward S, (1994), *Planning and Urban Change*, Paul Chapman Publishing, London.

World Commission on Environment and Development, (1987), *Our Common Future* (The Brundtland Report), Oxford University Press, Oxford.

Index

Printed and bound by CPI Group (UK) Ltd, Croydon, CR0 4YY

22/10/2024

01777626-0006